C-1921

CAREER EXAMINATION

THIS IS YOUR **PASSBOOK**® FOR ...

SAFETY COORDINATOR

NLC®

NATIONAL LEARNING CORPORATION®
passbooks.com

COPYRIGHT NOTICE

This book is SOLELY intended for, is sold ONLY to, and its use is RESTRICTED to individual, bona fide applicants or candidates who qualify by virtue of having seriously filed applications for appropriate license, certificate, professional and/or promotional advancement, higher school matriculation, scholarship, or other legitimate requirements of educational and/or governmental authorities.

This book is NOT intended for use, class instruction, tutoring, training, duplication, copying, reprinting, excerption, or adaptation, etc., by:

1) Other publishers
2) Proprietors and/or Instructors of «Coaching» and/or Preparatory Courses
3) Personnel and/or Training Divisions of commercial, industrial, and governmental organizations
4) Schools, colleges, or universities and/or their departments and staffs, including teachers and other personnel
5) Testing Agencies or Bureaus
6) Study groups which seek by the purchase of a single volume to copy and/or duplicate and/or adapt this material for use by the group as a whole without having purchased individual volumes for each of the members of the group
7) Et al.

Such persons would be in violation of appropriate Federal and State statutes.

PROVISION OF LICENSING AGREEMENTS. — Recognized educational, commercial, industrial, and governmental institutions and organizations, and others legitimately engaged in educational pursuits, including training, testing, and measurement activities, may address request for a licensing agreement to the copyright owners, who will determine whether, and under what conditions, including fees and charges, the materials in this book may be used them. In other words, a licensing facility exists for the legitimate use of the material in this book on other than an individual basis. However, it is asseverated and affirmed here that the material in this book CANNOT be used without the receipt of the express permission of such a licensing agreement from the Publishers. Inquiries re licensing should be addressed to the company, attention rights and permissions department.

All rights reserved, including the right of reproduction in whole or in part, in any form or by any means, electronic or mechanical, including photocopying, recording, or by any information storage and retrieval system, without permission in writing from the Publisher.

Copyright © 2022 by

NLC®

National Learning Corporation

212 Michael Drive, Syosset, NY 11791
(516) 921-8888 • www.passbooks.com
E-mail: info@passbooks.com

PUBLISHED IN THE UNITED STATES OF AMERICA

PASSBOOK® SERIES

THE *PASSBOOK® SERIES* has been created to prepare applicants and candidates for the ultimate academic battlefield – the examination room.

At some time in our lives, each and every one of us may be required to take an examination – for validation, matriculation, admission, qualification, registration, certification, or licensure.

Based on the assumption that every applicant or candidate has met the basic formal educational standards, has taken the required number of courses, and read the necessary texts, the *PASSBOOK® SERIES* furnishes the one special preparation which may assure passing with confidence, instead of failing with insecurity. Examination questions – together with answers – are furnished as the basic vehicle for study so that the mysteries of the examination and its compounding difficulties may be eliminated or diminished by a sure method.

This book is meant to help you pass your examination provided that you qualify and are serious in your objective.

The entire field is reviewed through the huge store of content information which is succinctly presented through a provocative and challenging approach – the question-and-answer method.

A climate of success is established by furnishing the correct answers at the end of each test.

You soon learn to recognize types of questions, forms of questions, and patterns of questioning. You may even begin to anticipate expected outcomes.

You perceive that many questions are repeated or adapted so that you can gain acute insights, which may enable you to score many sure points.

You learn how to confront new questions, or types of questions, and to attack them confidently and work out the correct answers.

You note objectives and emphases, and recognize pitfalls and dangers, so that you may make positive educational adjustments.

Moreover, you are kept fully informed in relation to new concepts, methods, practices, and directions in the field.

You discover that you arre actually taking the examination all the time: you are preparing for the examination by "taking" an examination, not by reading extraneous and/or supererogatory textbooks.

In short, this PASSBOOK®, used directedly, should be an important factor in helping you to pass your test.

SAFETY COORDINATOR

DUTIES

Directs and coordinates a safety program and is responsible for developing and implementing sound safety policies and practices. Responsible for investigating, reporting, and implementing a variety of rules and regulations established to protect employee health and safety, and for the public use of municipal owned property and facilities. The work is performed with wide latitude for the exercise of independent judgment. Does related work as required.

SUBJECT OF EXAMINATION

The written test will be designed to test for knowledge, skills, and/or abilities in such areas as:
1. Accident prevention and control;
2. Inspection, interviewing, and investigative techniques;
3. Statutory and regulatory requirements relating to occupational health and safety, and building safety;
4. Preparing written material;
5. Understanding and interpreting written material; and
6. Supervision; staff development and training.

HOW TO TAKE A TEST

I. YOU MUST PASS AN EXAMINATION

A. *WHAT EVERY CANDIDATE SHOULD KNOW*

Examination applicants often ask us for help in preparing for the written test. What can I study in advance? What kinds of questions will be asked? How will the test be given? How will the papers be graded?

As an applicant for a civil service examination, you may be wondering about some of these things. Our purpose here is to suggest effective methods of advance study and to describe civil service examinations.

Your chances for success on this examination can be increased if you know how to prepare. Those "pre-examination jitters" can be reduced if you know what to expect. You can even experience an adventure in good citizenship if you know why civil service exams are given.

B. *WHY ARE CIVIL SERVICE EXAMINATIONS GIVEN?*

Civil service examinations are important to you in two ways. As a citizen, you want public jobs filled by employees who know how to do their work. As a job seeker, you want a fair chance to compete for that job on an equal footing with other candidates. The best-known means of accomplishing this two-fold goal is the competitive examination.

Exams are widely publicized throughout the nation. They may be administered for jobs in federal, state, city, municipal, town or village governments or agencies.

Any citizen may apply, with some limitations, such as the age or residence of applicants. Your experience and education may be reviewed to see whether you meet the requirements for the particular examination. When these requirements exist, they are reasonable and applied consistently to all applicants. Thus, a competitive examination may cause you some uneasiness now, but it is your privilege and safeguard.

C. *HOW ARE CIVIL SERVICE EXAMS DEVELOPED?*

Examinations are carefully written by trained technicians who are specialists in the field known as "psychological measurement," in consultation with recognized authorities in the field of work that the test will cover. These experts recommend the subject matter areas or skills to be tested; only those knowledges or skills important to your success on the job are included. The most reliable books and source materials available are used as references. Together, the experts and technicians judge the difficulty level of the questions.

Test technicians know how to phrase questions so that the problem is clearly stated. Their ethics do not permit "trick" or "catch" questions. Questions may have been tried out on sample groups, or subjected to statistical analysis, to determine their usefulness.

Written tests are often used in combination with performance tests, ratings of training and experience, and oral interviews. All of these measures combine to form the best-known means of finding the right person for the right job.

II. HOW TO PASS THE WRITTEN TEST

A. NATURE OF THE EXAMINATION

To prepare intelligently for civil service examinations, you should know how they differ from school examinations you have taken. In school you were assigned certain definite pages to read or subjects to cover. The examination questions were quite detailed and usually emphasized memory. Civil service exams, on the other hand, try to discover your present ability to perform the duties of a position, plus your potentiality to learn these duties. In other words, a civil service exam attempts to predict how successful you will be. Questions cover such a broad area that they cannot be as minute and detailed as school exam questions.

In the public service similar kinds of work, or positions, are grouped together in one "class." This process is known as *position-classification*. All the positions in a class are paid according to the salary range for that class. One class title covers all of these positions, and they are all tested by the same examination.

B. FOUR BASIC STEPS

1) Study the announcement

How, then, can you know what subjects to study? Our best answer is: "Learn as much as possible about the class of positions for which you've applied." The exam will test the knowledge, skills and abilities needed to do the work.

Your most valuable source of information about the position you want is the official exam announcement. This announcement lists the training and experience qualifications. Check these standards and apply only if you come reasonably close to meeting them.

The brief description of the position in the examination announcement offers some clues to the subjects which will be tested. Think about the job itself. Review the duties in your mind. Can you perform them, or are there some in which you are rusty? Fill in the blank spots in your preparation.

Many jurisdictions preview the written test in the exam announcement by including a section called "Knowledge and Abilities Required," "Scope of the Examination," or some similar heading. Here you will find out specifically what fields will be tested.

2) Review your own background

Once you learn in general what the position is all about, and what you need to know to do the work, ask yourself which subjects you already know fairly well and which need improvement. You may wonder whether to concentrate on improving your strong areas or on building some background in your fields of weakness. When the announcement has specified "some knowledge" or "considerable knowledge," or has used adjectives like "beginning principles of..." or "advanced ... methods," you can get a clue as to the number and difficulty of questions to be asked in any given field. More questions, and hence broader coverage, would be included for those subjects which are more important in the work. Now weigh your strengths and weaknesses against the job requirements and prepare accordingly.

3) Determine the level of the position

Another way to tell how intensively you should prepare is to understand the level of the job for which you are applying. Is it the entering level? In other words, is this the position in which beginners in a field of work are hired? Or is it an intermediate or advanced level? Sometimes this is indicated by such words as "Junior" or "Senior" in the class title. Other jurisdictions use Roman numerals to designate the level – Clerk I, Clerk II, for example. The word "Supervisor" sometimes appears in the title. If the level is not indicated by the title, check the description of duties. Will you be working under very close supervision, or will you have responsibility for independent decisions in this work?

4) Choose appropriate study materials

Now that you know the subjects to be examined and the relative amount of each subject to be covered, you can choose suitable study materials. For beginning level jobs, or even advanced ones, if you have a pronounced weakness in some aspect of your training, read a modern, standard textbook in that field. Be sure it is up to date and has general coverage. Such books are normally available at your library, and the librarian will be glad to help you locate one. For entry-level positions, questions of appropriate difficulty are chosen – neither highly advanced questions, nor those too simple. Such questions require careful thought but not advanced training.

If the position for which you are applying is technical or advanced, you will read more advanced, specialized material. If you are already familiar with the basic principles of your field, elementary textbooks would waste your time. Concentrate on advanced textbooks and technical periodicals. Think through the concepts and review difficult problems in your field.

These are all general sources. You can get more ideas on your own initiative, following these leads. For example, training manuals and publications of the government agency which employs workers in your field can be useful, particularly for technical and professional positions. A letter or visit to the government department involved may result in more specific study suggestions, and certainly will provide you with a more definite idea of the exact nature of the position you are seeking.

III. KINDS OF TESTS

Tests are used for purposes other than measuring knowledge and ability to perform specified duties. For some positions, it is equally important to test ability to make adjustments to new situations or to profit from training. In others, basic mental abilities not dependent on information are essential. Questions which test these things may not appear as pertinent to the duties of the position as those which test for knowledge and information. Yet they are often highly important parts of a fair examination. For very general questions, it is almost impossible to help you direct your study efforts. What we can do is to point out some of the more common of these general abilities needed in public service positions and describe some typical questions.

1) General information

Broad, general information has been found useful for predicting job success in some kinds of work. This is tested in a variety of ways, from vocabulary lists to questions about current events. Basic background in some field of work, such as sociology or economics, may be sampled in a group of questions. Often these are principles which have become familiar to most persons through exposure rather than through formal training. It is difficult to advise you how to study for these questions; being alert to the world around you is our best suggestion.

2) Verbal ability

An example of an ability needed in many positions is verbal or language ability. Verbal ability is, in brief, the ability to use and understand words. Vocabulary and grammar tests are typical measures of this ability. Reading comprehension or paragraph interpretation questions are common in many kinds of civil service tests. You are given a paragraph of written material and asked to find its central meaning.

3) Numerical ability

Number skills can be tested by the familiar arithmetic problem, by checking paired lists of numbers to see which are alike and which are different, or by interpreting charts and graphs. In the latter test, a graph may be printed in the test booklet which you are asked to use as the basis for answering questions.

4) Observation

A popular test for law-enforcement positions is the observation test. A picture is shown to you for several minutes, then taken away. Questions about the picture test your ability to observe both details and larger elements.

5) Following directions

In many positions in the public service, the employee must be able to carry out written instructions dependably and accurately. You may be given a chart with several columns, each column listing a variety of information. The questions require you to carry out directions involving the information given in the chart.

6) Skills and aptitudes

Performance tests effectively measure some manual skills and aptitudes. When the skill is one in which you are trained, such as typing or shorthand, you can practice. These tests are often very much like those given in business school or high school courses. For many of the other skills and aptitudes, however, no short-time preparation can be made. Skills and abilities natural to you or that you have developed throughout your lifetime are being tested.

Many of the general questions just described provide all the data needed to answer the questions and ask you to use your reasoning ability to find the answers. Your best preparation for these tests, as well as for tests of facts and ideas, is to be at your physical and mental best. You, no doubt, have your own methods of getting into an exam-taking mood and keeping "in shape." The next section lists some ideas on this subject.

IV. KINDS OF QUESTIONS

Only rarely is the "essay" question, which you answer in narrative form, used in civil service tests. Civil service tests are usually of the short-answer type. Full instructions for answering these questions will be given to you at the examination. But in case this is your first experience with short-answer questions and separate answer sheets, here is what you need to know:

1) Multiple-choice Questions

Most popular of the short-answer questions is the "multiple choice" or "best answer" question. It can be used, for example, to test for factual knowledge, ability to solve problems or judgment in meeting situations found at work.

A multiple-choice question is normally one of three types—
- It can begin with an incomplete statement followed by several possible endings. You are to find the one ending which *best* completes the statement, although some of the others may not be entirely wrong.
- It can also be a complete statement in the form of a question which is answered by choosing one of the statements listed.
- It can be in the form of a problem – again you select the best answer.

Here is an example of a multiple-choice question with a discussion which should give you some clues as to the method for choosing the right answer:

When an employee has a complaint about his assignment, the action which will *best* help him overcome his difficulty is to
  A. discuss his difficulty with his coworkers
  B. take the problem to the head of the organization
  C. take the problem to the person who gave him the assignment
  D. say nothing to anyone about his complaint

In answering this question, you should study each of the choices to find which is best. Consider choice "A" – Certainly an employee may discuss his complaint with fellow employees, but no change or improvement can result, and the complaint remains unresolved. Choice "B" is a poor choice since the head of the organization probably does not know what assignment you have been given, and taking your problem to him is known as "going over the head" of the supervisor. The supervisor, or person who made the assignment, is the person who can clarify it or correct any injustice. Choice "C" is, therefore, correct. To say nothing, as in choice "D," is unwise. Supervisors have and interest in knowing the problems employees are facing, and the employee is seeking a solution to his problem.

2) True/False Questions

The "true/false" or "right/wrong" form of question is sometimes used. Here a complete statement is given. Your job is to decide whether the statement is right or wrong.

SAMPLE: A roaming cell-phone call to a nearby city costs less than a non-roaming call to a distant city.

This statement is wrong, or false, since roaming calls are more expensive.
This is *not* a complete list of all possible question forms, although most of the others are variations of these common types. You will always get complete directions for

answering questions. Be sure you understand *how* to mark your answers – ask questions until you do.

V. RECORDING YOUR ANSWERS

Computer terminals are used more and more today for many different kinds of exams.

For an examination with very few applicants, you may be told to record your answers in the test booklet itself. Separate answer sheets are much more common. If this separate answer sheet is to be scored by machine – and this is often the case – it is highly important that you mark your answers correctly in order to get credit.

An electronic scoring machine is often used in civil service offices because of the speed with which papers can be scored. Machine-scored answer sheets must be marked with a pencil, which will be given to you. This pencil has a high graphite content which responds to the electronic scoring machine. As a matter of fact, stray dots may register as answers, so do not let your pencil rest on the answer sheet while you are pondering the correct answer. Also, if your pencil lead breaks or is otherwise defective, ask for another.

Since the answer sheet will be dropped in a slot in the scoring machine, be careful not to bend the corners or get the paper crumpled.

The answer sheet normally has five vertical columns of numbers, with 30 numbers to a column. These numbers correspond to the question numbers in your test booklet. After each number, going across the page are four or five pairs of dotted lines. These short dotted lines have small letters or numbers above them. The first two pairs may also have a "T" or "F" above the letters. This indicates that the first two pairs only are to be used if the questions are of the true-false type. If the questions are multiple choice, disregard the "T" and "F" and pay attention only to the small letters or numbers.

Answer your questions in the manner of the sample that follows:

32. The largest city in the United States is
   A. Washington, D.C.
   B. New York City
   C. Chicago
   D. Detroit
   E. San Francisco

1) Choose the answer you think is best. (New York City is the largest, so "B" is correct.)
2) Find the row of dotted lines numbered the same as the question you are answering. (Find row number 32)
3) Find the pair of dotted lines corresponding to the answer. (Find the pair of lines under the mark "B.")
4) Make a solid black mark between the dotted lines.

VI. BEFORE THE TEST

Common sense will help you find procedures to follow to get ready for an examination. Too many of us, however, overlook these sensible measures. Indeed, nervousness and fatigue have been found to be the most serious reasons why applicants fail to do their best on civil service tests. Here is a list of reminders:

- Begin your preparation early – Don't wait until the last minute to go scurrying around for books and materials or to find out what the position is all about.
- Prepare continuously – An hour a night for a week is better than an all-night cram session. This has been definitely established. What is more, a night a week for a month will return better dividends than crowding your study into a shorter period of time.
- Locate the place of the exam – You have been sent a notice telling you when and where to report for the examination. If the location is in a different town or otherwise unfamiliar to you, it would be well to inquire the best route and learn something about the building.
- Relax the night before the test – Allow your mind to rest. Do not study at all that night. Plan some mild recreation or diversion; then go to bed early and get a good night's sleep.
- Get up early enough to make a leisurely trip to the place for the test – This way unforeseen events, traffic snarls, unfamiliar buildings, etc. will not upset you.
- Dress comfortably – A written test is not a fashion show. You will be known by number and not by name, so wear something comfortable.
- Leave excess paraphernalia at home – Shopping bags and odd bundles will get in your way. You need bring only the items mentioned in the official notice you received; usually everything you need is provided. Do not bring reference books to the exam. They will only confuse those last minutes and be taken away from you when in the test room.
- Arrive somewhat ahead of time – If because of transportation schedules you must get there very early, bring a newspaper or magazine to take your mind off yourself while waiting.
- Locate the examination room – When you have found the proper room, you will be directed to the seat or part of the room where you will sit. Sometimes you are given a sheet of instructions to read while you are waiting. Do not fill out any forms until you are told to do so; just read them and be prepared.
- Relax and prepare to listen to the instructions
- If you have any physical problem that may keep you from doing your best, be sure to tell the test administrator. If you are sick or in poor health, you really cannot do your best on the exam. You can come back and take the test some other time.

VII. AT THE TEST

The day of the test is here and you have the test booklet in your hand. The temptation to get going is very strong. Caution! There is more to success than knowing the right answers. You must know how to identify your papers and understand variations in the type of short-answer question used in this particular examination. Follow these suggestions for maximum results from your efforts:

1) Cooperate with the monitor

The test administrator has a duty to create a situation in which you can be as much at ease as possible. He will give instructions, tell you when to begin, check to see that you are marking your answer sheet correctly, and so on. He is not there to guard you, although he will see that your competitors do not take unfair advantage. He wants to help you do your best.

2) Listen to all instructions

Don't jump the gun! Wait until you understand all directions. In most civil service tests you get more time than you need to answer the questions. So don't be in a hurry. Read each word of instructions until you clearly understand the meaning. Study the examples, listen to all announcements and follow directions. Ask questions if you do not understand what to do.

3) Identify your papers

Civil service exams are usually identified by number only. You will be assigned a number; you must not put your name on your test papers. Be sure to copy your number correctly. Since more than one exam may be given, copy your exact examination title.

4) Plan your time

Unless you are told that a test is a "speed" or "rate of work" test, speed itself is usually not important. Time enough to answer all the questions will be provided, but this does not mean that you have all day. An overall time limit has been set. Divide the total time (in minutes) by the number of questions to determine the approximate time you have for each question.

5) Do not linger over difficult questions

If you come across a difficult question, mark it with a paper clip (useful to have along) and come back to it when you have been through the booklet. One caution if you do this – be sure to skip a number on your answer sheet as well. Check often to be sure that you have not lost your place and that you are marking in the row numbered the same as the question you are answering.

6) Read the questions

Be sure you know what the question asks! Many capable people are unsuccessful because they failed to *read* the questions correctly.

7) Answer all questions

Unless you have been instructed that a penalty will be deducted for incorrect answers, it is better to guess than to omit a question.

8) Speed tests

It is often better NOT to guess on speed tests. It has been found that on timed tests people are tempted to spend the last few seconds before time is called in marking answers at random – without even reading them – in the hope of picking up a few extra points. To discourage this practice, the instructions may warn you that your score will be "corrected" for guessing. That is, a penalty will be applied. The incorrect answers will be deducted from the correct ones, or some other penalty formula will be used.

9) Review your answers

If you finish before time is called, go back to the questions you guessed or omitted to give them further thought. Review other answers if you have time.

10) Return your test materials

If you are ready to leave before others have finished or time is called, take ALL your materials to the monitor and leave quietly. Never take any test material with you. The monitor can discover whose papers are not complete, and taking a test booklet may be grounds for disqualification.

VIII. EXAMINATION TECHNIQUES

1) Read the general instructions carefully. These are usually printed on the first page of the exam booklet. As a rule, these instructions refer to the timing of the examination; the fact that you should not start work until the signal and must stop work at a signal, etc. If there are any *special* instructions, such as a choice of questions to be answered, make sure that you note this instruction carefully.

2) When you are ready to start work on the examination, that is as soon as the signal has been given, read the instructions to each question booklet, underline any key words or phrases, such as *least, best, outline, describe* and the like. In this way you will tend to answer as requested rather than discover on reviewing your paper that you *listed without describing*, that you selected the *worst* choice rather than the *best* choice, etc.

3) If the examination is of the objective or multiple-choice type – that is, each question will also give a series of possible answers: A, B, C or D, and you are called upon to select the best answer and write the letter next to that answer on your answer paper – it is advisable to start answering each question in turn. There may be anywhere from 50 to 100 such questions in the three or four hours allotted and you can see how much time would be taken if you read through all the questions before beginning to answer any. Furthermore, if you come across a question or group of questions which you know would be difficult to answer, it would undoubtedly affect your handling of all the other questions.

4) If the examination is of the essay type and contains but a few questions, it is a moot point as to whether you should read all the questions before starting to answer any one. Of course, if you are given a choice – say five out of seven and the like – then it is essential to read all the questions so you can eliminate the two that are most difficult. If, however, you are asked to answer all the questions, there may be danger in trying to answer the easiest one first because you may find that you will spend too much time on it. The best technique is to answer the first question, then proceed to the second, etc.

5) Time your answers. Before the exam begins, write down the time it started, then add the time allowed for the examination and write down the time it must be completed, then divide the time available somewhat as follows:

- If 3-1/2 hours are allowed, that would be 210 minutes. If you have 80 objective-type questions, that would be an average of 2-1/2 minutes per question. Allow yourself no more than 2 minutes per question, or a total of 160 minutes, which will permit about 50 minutes to review.
- If for the time allotment of 210 minutes there are 7 essay questions to answer, that would average about 30 minutes a question. Give yourself only 25 minutes per question so that you have about 35 minutes to review.

6) The most important instruction is to *read each question* and make sure you know what is wanted. The second most important instruction is to *time yourself properly* so that you answer every question. The third most important instruction is to *answer every question*. Guess if you have to but include something for each question. Remember that you will receive no credit for a blank and will probably receive some credit if you write something in answer to an essay question. If you guess a letter – say "B" for a multiple-choice question – you may have guessed right. If you leave a blank as an answer to a multiple-choice question, the examiners may respect your feelings but it will not add a point to your score. Some exams may penalize you for wrong answers, so in such cases *only*, you may not want to guess unless you have some basis for your answer.

7) Suggestions
   a. Objective-type questions
      1. Examine the question booklet for proper sequence of pages and questions
      2. Read all instructions carefully
      3. Skip any question which seems too difficult; return to it after all other questions have been answered
      4. Apportion your time properly; do not spend too much time on any single question or group of questions
      5. Note and underline key words – *all, most, fewest, least, best, worst, same, opposite,* etc.
      6. Pay particular attention to negatives
      7. Note unusual option, e.g., unduly long, short, complex, different or similar in content to the body of the question
      8. Observe the use of "hedging" words – *probably, may, most likely,* etc.
      9. Make sure that your answer is put next to the same number as the question
      10. Do not second-guess unless you have good reason to believe the second answer is definitely more correct
      11. Cross out original answer if you decide another answer is more accurate; do not erase until you are ready to hand your paper in
      12. Answer all questions; guess unless instructed otherwise
      13. Leave time for review

   b. Essay questions
      1. Read each question carefully
      2. Determine exactly what is wanted. Underline key words or phrases.
      3. Decide on outline or paragraph answer

4. Include many different points and elements unless asked to develop any one or two points or elements
5. Show impartiality by giving pros and cons unless directed to select one side only
6. Make and write down any assumptions you find necessary to answer the questions
7. Watch your English, grammar, punctuation and choice of words
8. Time your answers; don't crowd material

8) Answering the essay question

Most essay questions can be answered by framing the specific response around several key words or ideas. Here are a few such key words or ideas:

M's: manpower, materials, methods, money, management
P's: purpose, program, policy, plan, procedure, practice, problems, pitfalls, personnel, public relations

   a. Six basic steps in handling problems:
      1. Preliminary plan and background development
      2. Collect information, data and facts
      3. Analyze and interpret information, data and facts
      4. Analyze and develop solutions as well as make recommendations
      5. Prepare report and sell recommendations
      6. Install recommendations and follow up effectiveness

   b. Pitfalls to avoid
      1. *Taking things for granted* – A statement of the situation does not necessarily imply that each of the elements is necessarily true; for example, a complaint may be invalid and biased so that all that can be taken for granted is that a complaint has been registered
      2. *Considering only one side of a situation* – Wherever possible, indicate several alternatives and then point out the reasons you selected the best one
      3. *Failing to indicate follow up* – Whenever your answer indicates action on your part, make certain that you will take proper follow-up action to see how successful your recommendations, procedures or actions turn out to be
      4. *Taking too long in answering any single question* – Remember to time your answers properly

IX. AFTER THE TEST

Scoring procedures differ in detail among civil service jurisdictions although the general principles are the same. Whether the papers are hand-scored or graded by machine we have described, they are nearly always graded by number. That is, the person who marks the paper knows only the number – never the name – of the applicant. Not until all the papers have been graded will they be matched with names. If other tests, such as training and experience or oral interview ratings have been given, scores will be combined. Different parts of the examination usually have different weights. For example, the written test might count 60 percent of the final grade, and a rating of training and experience 40 percent. In many jurisdictions, veterans will have a certain number of points added to their grades.

After the final grade has been determined, the names are placed in grade order and an eligible list is established. There are various methods for resolving ties between those who get the same final grade – probably the most common is to place first the name of the person whose application was received first. Job offers are made from the eligible list in the order the names appear on it. You will be notified of your grade and your rank as soon as all these computations have been made. This will be done as rapidly as possible.

People who are found to meet the requirements in the announcement are called "eligibles." Their names are put on a list of eligible candidates. An eligible's chances of getting a job depend on how high he stands on this list and how fast agencies are filling jobs from the list.

When a job is to be filled from a list of eligibles, the agency asks for the names of people on the list of eligibles for that job. When the civil service commission receives this request, it sends to the agency the names of the three people highest on this list. Or, if the job to be filled has specialized requirements, the office sends the agency the names of the top three persons who meet these requirements from the general list.

The appointing officer makes a choice from among the three people whose names were sent to him. If the selected person accepts the appointment, the names of the others are put back on the list to be considered for future openings.

That is the rule in hiring from all kinds of eligible lists, whether they are for typist, carpenter, chemist, or something else. For every vacancy, the appointing officer has his choice of any one of the top three eligibles on the list. This explains why the person whose name is on top of the list sometimes does not get an appointment when some of the persons lower on the list do. If the appointing officer chooses the second or third eligible, the No. 1 eligible does not get a job at once, but stays on the list until he is appointed or the list is terminated.

X. HOW TO PASS THE INTERVIEW TEST

The examination for which you applied requires an oral interview test. You have already taken the written test and you are now being called for the interview test – the final part of the formal examination.

You may think that it is not possible to prepare for an interview test and that there are no procedures to follow during an interview. Our purpose is to point out some things you can do in advance that will help you and some good rules to follow and pitfalls to avoid while you are being interviewed.

What is an interview supposed to test?
The written examination is designed to test the technical knowledge and competence of the candidate; the oral is designed to evaluate intangible qualities, not readily measured otherwise, and to establish a list showing the relative fitness of each candidate – as measured against his competitors – for the position sought. Scoring is not on the basis of "right" and "wrong," but on a sliding scale of values ranging from "not passable" to "outstanding." As a matter of fact, it is possible to achieve a relatively low score without a single "incorrect" answer because of evident weakness in the qualities being measured.

Occasionally, an examination may consist entirely of an oral test – either an individual or a group oral. In such cases, information is sought concerning the technical knowledges and abilities of the candidate, since there has been no written examination for this purpose. More commonly, however, an oral test is used to supplement a written examination.

Who conducts interviews?

The composition of oral boards varies among different jurisdictions. In nearly all, a representative of the personnel department serves as chairman. One of the members of the board may be a representative of the department in which the candidate would work. In some cases, "outside experts" are used, and, frequently, a businessman or some other representative of the general public is asked to serve. Labor and management or other special groups may be represented. The aim is to secure the services of experts in the appropriate field.

However the board is composed, it is a good idea (and not at all improper or unethical) to ascertain in advance of the interview who the members are and what groups they represent. When you are introduced to them, you will have some idea of their backgrounds and interests, and at least you will not stutter and stammer over their names.

What should be done before the interview?

While knowledge about the board members is useful and takes some of the surprise element out of the interview, there is other preparation which is more substantive. It *is* possible to prepare for an oral interview – in several ways:

1) Keep a copy of your application and review it carefully before the interview

This may be the only document before the oral board, and the starting point of the interview. Know what education and experience you have listed there, and the sequence and dates of all of it. Sometimes the board will ask you to review the highlights of your experience for them; you should not have to hem and haw doing it.

2) Study the class specification and the examination announcement

Usually, the oral board has one or both of these to guide them. The qualities, characteristics or knowledges required by the position sought are stated in these documents. They offer valuable clues as to the nature of the oral interview. For example, if the job involves supervisory responsibilities, the announcement will usually indicate that knowledge of modern supervisory methods and the qualifications of the candidate as a supervisor will be tested. If so, you can expect such questions, frequently in the form of a hypothetical situation which you are expected to solve. NEVER go into an oral without knowledge of the duties and responsibilities of the job you seek.

3) Think through each qualification required

Try to visualize the kind of questions you would ask if you were a board member. How well could you answer them? Try especially to appraise your own knowledge and background in each area, *measured against the job sought*, and identify any areas in which you are weak. Be critical and realistic – do not flatter yourself.

4) Do some general reading in areas in which you feel you may be weak

For example, if the job involves supervision and your past experience has NOT, some general reading in supervisory methods and practices, particularly in the field of human relations, might be useful. Do NOT study agency procedures or detailed manuals. The oral board will be testing your understanding and capacity, not your memory.

5) Get a good night's sleep and watch your general health and mental attitude

You will want a clear head at the interview. Take care of a cold or any other minor ailment, and of course, no hangovers.

What should be done on the day of the interview?

Now comes the day of the interview itself. Give yourself plenty of time to get there. Plan to arrive somewhat ahead of the scheduled time, particularly if your appointment is in the fore part of the day. If a previous candidate fails to appear, the board might be ready for you a bit early. By early afternoon an oral board is almost invariably behind schedule if there are many candidates, and you may have to wait. Take along a book or magazine to read, or your application to review, but leave any extraneous material in the waiting room when you go in for your interview. In any event, relax and compose yourself.

The matter of dress is important. The board is forming impressions about you – from your experience, your manners, your attitude, and your appearance. Give your personal appearance careful attention. Dress your best, but not your flashiest. Choose conservative, appropriate clothing, and be sure it is immaculate. This is a business interview, and your appearance should indicate that you regard it as such. Besides, being well groomed and properly dressed will help boost your confidence.

Sooner or later, someone will call your name and escort you into the interview room. *This is it.* From here on you are on your own. It is too late for any more preparation. But remember, you asked for this opportunity to prove your fitness, and you are here because your request was granted.

What happens when you go in?

The usual sequence of events will be as follows: The clerk (who is often the board stenographer) will introduce you to the chairman of the oral board, who will introduce you to the other members of the board. Acknowledge the introductions before you sit down. Do not be surprised if you find a microphone facing you or a stenotypist sitting by. Oral interviews are usually recorded in the event of an appeal or other review.

Usually the chairman of the board will open the interview by reviewing the highlights of your education and work experience from your application – primarily for the benefit of the other members of the board, as well as to get the material into the record. Do not interrupt or comment unless there is an error or significant misinterpretation; if that is the case, do not hesitate. But do not quibble about insignificant matters. Also, he will usually ask you some question about your education, experience or your present job – partly to get you to start talking and to establish the interviewing "rapport." He may start the actual questioning, or turn it over to one of the other members. Frequently, each member undertakes the questioning on a particular area, one in which he is perhaps most competent, so you can expect each member to participate in the examination. Because time is limited, you may also expect some rather abrupt switches in the direction the questioning takes, so do not be upset by it. Normally, a board member will not pursue a single line of questioning unless he discovers a particular strength or weakness.

After each member has participated, the chairman will usually ask whether any member has any further questions, then will ask you if you have anything you wish to add. Unless you are expecting this question, it may floor you. Worse, it may start you off on an extended, extemporaneous speech. The board is not usually seeking more information. The question is principally to offer you a last opportunity to present further qualifications or to indicate that you have nothing to add. So, if you feel that a significant qualification or characteristic has been overlooked, it is proper to point it out in a sentence or so. Do not compliment the board on the thoroughness of their examination – they have been sketchy, and you know it. If you wish, merely say, "No thank you, I have nothing further to add." This is a point where you can "talk yourself out" of a good impression or fail to present an important bit of information. Remember, *you close the interview yourself.*

The chairman will then say, "That is all, Mr. _____, thank you." Do not be startled; the interview is over, and quicker than you think. Thank him, gather your belongings and take your leave. Save your sigh of relief for the other side of the door.

How to put your best foot forward

Throughout this entire process, you may feel that the board individually and collectively is trying to pierce your defenses, seek out your hidden weaknesses and embarrass and confuse you. Actually, this is not true. They are obliged to make an appraisal of your qualifications for the job you are seeking, and they want to see you in your best light. Remember, they must interview all candidates and a non-cooperative candidate may become a failure in spite of their best efforts to bring out his qualifications. Here are 15 suggestions that will help you:

1) Be natural – Keep your attitude confident, not cocky

If you are not confident that you can do the job, do not expect the board to be. Do not apologize for your weaknesses, try to bring out your strong points. The board is interested in a positive, not negative, presentation. Cockiness will antagonize any board member and make him wonder if you are covering up a weakness by a false show of strength.

2) Get comfortable, but don't lounge or sprawl

Sit erectly but not stiffly. A careless posture may lead the board to conclude that you are careless in other things, or at least that you are not impressed by the importance of the occasion. Either conclusion is natural, even if incorrect. Do not fuss with your clothing, a pencil or an ashtray. Your hands may occasionally be useful to emphasize a point; do not let them become a point of distraction.

3) Do not wisecrack or make small talk

This is a serious situation, and your attitude should show that you consider it as such. Further, the time of the board is limited – they do not want to waste it, and neither should you.

4) Do not exaggerate your experience or abilities

In the first place, from information in the application or other interviews and sources, the board may know more about you than you think. Secondly, you probably will not get away with it. An experienced board is rather adept at spotting such a situation, so do not take the chance.

5) If you know a board member, do not make a point of it, yet do not hide it

Certainly you are not fooling him, and probably not the other members of the board. Do not try to take advantage of your acquaintanceship – it will probably do you little good.

6) Do not dominate the interview

Let the board do that. They will give you the clues – do not assume that you have to do all the talking. Realize that the board has a number of questions to ask you, and do not try to take up all the interview time by showing off your extensive knowledge of the answer to the first one.

7) Be attentive

You only have 20 minutes or so, and you should keep your attention at its sharpest throughout. When a member is addressing a problem or question to you, give him your undivided attention. Address your reply principally to him, but do not exclude the other board members.

8) Do not interrupt

A board member may be stating a problem for you to analyze. He will ask you a question when the time comes. Let him state the problem, and wait for the question.

9) Make sure you understand the question

Do not try to answer until you are sure what the question is. If it is not clear, restate it in your own words or ask the board member to clarify it for you. However, do not haggle about minor elements.

10) Reply promptly but not hastily

A common entry on oral board rating sheets is "candidate responded readily," or "candidate hesitated in replies." Respond as promptly and quickly as you can, but do not jump to a hasty, ill-considered answer.

11) Do not be peremptory in your answers

A brief answer is proper – but do not fire your answer back. That is a losing game from your point of view. The board member can probably ask questions much faster than you can answer them.

12) Do not try to create the answer you think the board member wants

He is interested in what kind of mind you have and how it works – not in playing games. Furthermore, he can usually spot this practice and will actually grade you down on it.

13) Do not switch sides in your reply merely to agree with a board member

Frequently, a member will take a contrary position merely to draw you out and to see if you are willing and able to defend your point of view. Do not start a debate, yet do not surrender a good position. If a position is worth taking, it is worth defending.

14) Do not be afraid to admit an error in judgment if you are shown to be wrong

The board knows that you are forced to reply without any opportunity for careful consideration. Your answer may be demonstrably wrong. If so, admit it and get on with the interview.

15) Do not dwell at length on your present job

The opening question may relate to your present assignment. Answer the question but do not go into an extended discussion. You are being examined for a *new* job, not your present one. As a matter of fact, try to phrase ALL your answers in terms of the job for which you are being examined.

Basis of Rating

Probably you will forget most of these "do's" and "don'ts" when you walk into the oral interview room. Even remembering them all will not ensure you a passing grade. Perhaps you did not have the qualifications in the first place. But remembering them will help you to put your best foot forward, without treading on the toes of the board members.

Rumor and popular opinion to the contrary notwithstanding, an oral board wants you to make the best appearance possible. They know you are under pressure – but they also want to see how you respond to it as a guide to what your reaction would be under the pressures of the job you seek. They will be influenced by the degree of poise you display, the personal traits you show and the manner in which you respond.

ABOUT THIS BOOK

This book contains tests divided into Examination Sections. Go through each test, answering every question in the margin. At the end of each test look at the answer key and check your answers. On the ones you got wrong, look at the right answer choice and learn. Do not fill in the answers first. Do not memorize the questions and answers, but understand the answer and principles involved. On your test, the questions will likely be different from the samples. Questions are changed and new ones added. If you understand these past questions you should have success with any changes that arise. Tests may consist of several types of questions. We have additional books on each subject should more study be advisable or necessary for you. Finally, the more you study, the better prepared you will be. This book is intended to be the last thing you study before you walk into the examination room. Prior study of relevant texts is also recommended. NLC publishes some of these in our Fundamental Series. Knowledge and good sense are important factors in passing your exam. Good luck also helps. So now study this Passbook, absorb the material contained within and take that knowledge into the examination. Then do your best to pass that exam.

EXAMINATION SECTION

SAFETY
EXAMINATION SECTION
TEST 1

DIRECTIONS: Each question or incomplete statement is followed by several suggested answers or completions. Select the one that BEST answers the question or completes the statement. *PRINT THE LETTER OF THE CORRECT ANSWER IN THE SPACE AT THE RIGHT.*

1. Which one of the following is an INCORRECT safety guideline? 1._____

   A. All working conditions and equipment should be considered carefully before beginning an operation.
   B. Aisles should be lighted properly.
   C. Personnel should be provided with protective clothing essential to safe performance of a task.
   D. In manual lifting, the worker must keep his knees straight and lift with the arm muscles.

2. Of the following, the supply item with the GREATEST susceptibility to spontaneous heating is 2._____

   A. alcohol, ethyl   B. kerosene
   C. candles   D. turpentine

Questions 3-7.

DIRECTIONS: Questions 3 through 7 are descriptions of accidents that occurred in a warehouse. For each accident, choose the letter in front of the safety measure that is MOST likely to prevent a repetition of the accident indicated.

SAFETY MEASURE

   A. Posting warning signs
   B. Redesign of layout or facilities
   C. Repairing, improving or replacing supplies, tools or equipment
   D. Training the staff in safe practices

3. After a new all-glass door was installed at the entrance to the warehouse, one of the employees banged his head into the door causing a large lump on his forehead when he failed to realize that the door was closed. 3._____

4. While tieing up a package with manila rope, an employee got several small rope splinters in his right hand and he had to have medical treatment to remove the splinters. 4._____

5. An employee discovered a small fire in a wastepaper basket but was unable to prevent it from spreading because all the nearby fire extinguishers were inaccessible due to skids of material being stacked in front of the extinguishers. 5._____

6. When a laborer attempted to drop the tailgate of a delivery truck while the truck was being backed into the loading dock, he had his fingers crushed when the truck continued to move while he was working on lowering the tailgate. 6._____

1

2 (#1)

7. An employee carrying a carton with both hands tripped over a broom which had been left lying in an aisle by another employee after the latter had swept the aisle. 7._____

8. Safety experts agree that accidents can probably BEST be prevented by 8._____

    A. developing safety consciousness among employees
    B. developing a program which publicizes major accidents
    C. penalizing employees the first time they do not follow safety procedures
    D. giving recognition to employees with accident-free records

9. The accident records of many agencies indicate that most on-the-job injuries are caused by the unsafe acts of their employees.
   Which one of the following statements pinpoints the MOST probable cause of this safety problem? 9._____

    A. Responsibility for preventing on-the-job accidents has not been delegated.
    B. Lack of proper supervision has permitted these unsafe actions to continue.
    C. No consideration has been given to eliminating environmental job hazards.
    D. Penalties for causing on-the-job accidents are not sufficiently severe.

10. Which of the following methods is LEAST essential to the success of an accident prevention program? 10._____

    A. Determining corrective measures by analyzing the causes of accidents and making recommendations to eliminate them
    B. Educating employees as to the importance of safe working conditions and methods
    C. Determining accident causes by seeking out the conditions from which each accident has developed
    D. Holding each supervisor responsible for accidents occurring during the on-the-job performance of his immediate subordinates

11. The effectiveness of a public relations program in a public agency is BEST indicated by the 11._____

    A. amount of mass media publicity favorable to the policies of the agency
    B. morale of those employees who directly serve the patrons of the agency
    C. public's understanding and support of the agency's program and policies
    D. number of complaints received by the agency from patrons using its facilities

12. Buttered bread and coffee dropped on an office floor in a terminal are 12._____

    A. minor hazards which should cause no serious injury
    B. unattractive, but not dangerous
    C. the most dangerous types of office hazards
    D. hazards which should be corrected immediately

13. A laborer was sent upstairs to get a 20-pound sack of rock salt. While going downstairs and reading the printing on the sack, he fell, and the sack of rock salt fell and broke his toe.
    Which of the following is MOST likely to have been the MOST important cause of the accident?
    The 13._____

A. stairs were beginning to become worn
B. laborer was carrying too heavy a sack of rock salt
C. rock salt was in a place that was too inaccessible
D. laborer was not careful about the way he went down the stairs

14. A COMMONLY recommended safe distance between the foot of an extension ladder and the wall against which it is placed is

A. 3 feet for ladders less than 18 feet in height
B. between 3 feet and 6 feet for ladders less than 18 feet in length
C. 1/8 the length of the extended ladder
D. 1/4 the length of the extended ladder

15. The BEST type of fire extinguisher for electrical fires is the _____ extinguisher.

A. dry chemical
B. foam
C. carbon monoxide
D. baking soda-acid

16. A Class A extinguisher should be used for fires in

A. potassium, magnesium, zinc, sodium
B. electrical wiring
C. oil, gasoline
D. wood, paper, and textiles

17. The one of the following which is NOT a safe practice when lifting heavy objects is:

A. Keep the back as nearly upright as possible
B. If the object feels too heavy, keep lifting until you get help
C. Spread the feet apart
D. Use the arm and leg muscles

18. In a shop, it would be MOST necessary to provide a fitted cover on the metal container for

A. old paint brushes
B. oily rags and waste
C. sand
D. broken glass

19. Safety shoes usually have the unique feature of

A. extra hard heels and soles to prevent nails from piercing the shoes
B. special leather to prevent the piercing of the shoes by falling objects
C. a metal guard over the toes which is built into the shoes
D. a non-slip tread on the heels and soles

20. Of the following, the MOST important factor contributing to a helper's safety on the job is for him to

A. work slowly
B. wear gloves
C. be alert
D. know his job well

21. If it is necessary for you to lift one end of a piece of heavy equipment with a crowbar in order to allow a maintainer to work underneath it, the BEST of the following procedures to follow is to

    A. support the handle of the bar on a box
    B. insert temporary blocks to support the piece
    C. call the supervisor to help you
    D. wear heavy gloves

22. Of the following, the MOST important reason for not letting oily rags accumulate in an open storage bin is that they

    A. may start a fire by spontaneous combustion
    B. will drip oil onto other items in the bin
    C. may cause a foul odor
    D. will make the area messy

23. Of the following, the BEST method to employ in putting out a gasoline fire is to

    A. use a bucket of water
    B. smother it with rags
    C. use a carbon dioxide extinguisher
    D. use a carbon tetrachloride extinguisher

24. When opening an emergency exit door set in the sidewalk, the door should be raised slowly to avoid

    A. a sudden rush of air from the street
    B. making unnecessary noise
    C. damage to the sidewalk
    D. injuring pedestrians

25. The BEST reason to turn off lights when cleaning lampshades on electrical fixtures is to

    A. conserve energy
    B. avoid electrical shock
    C. prevent breakage of lightbulbs
    D. prevent unnecessary eye strain

KEY (CORRECT ANSWERS)

1.	D	11.	C
2.	D	12.	D
3.	A	13.	D
4.	D	14.	D
5.	B	15.	A
6.	D	16.	D
7.	D	17.	B
8.	A	18.	B
9.	B	19.	C
10.	D	20.	C

21. B
22. A
23. C
24. D
25. B

TEST 2

DIRECTIONS: Each question or incomplete statement is followed by several suggested answers or completions. Select the one that BEST answers the question or completes the statement. *PRINT THE LETTER OF THE CORRECT ANSWER IN THE SPACE AT THE RIGHT.*

1. The MOST important reason for roping off a work area in a terminal is to

   A. protect the public
   B. protect the repair crew
   C. prevent distraction of the crew by the public
   D. prevent delays to the public

2. Shoes which have a sponge rubber sole should NOT be worn around a work area because such a sole

   A. will wear quickly
   B. is not waterproof
   C. does not keep the feet warm
   D. is easily punctured by steel objects

3. When repair work is being done on an elevated structure, canvas spreads are suspended under the working area MAINLY to

   A. reduce noise
   B. discourage crowds
   C. protect the structure
   D. protect pedestrians

4. It is poor practice to hold a piece of wood in the hands or lap when tightening a screw in the wood.
   This is for the reason that

   A. sufficient leverage cannot be obtained
   B. the screwdriver may bend
   C. the wood will probably split
   D. personal injury is likely to result

5. Steel helmets give workers the MOST protection from

   A. falling objects
   B. eye injuries
   C. fire
   D. electric shock

6. It is POOR practice to wear goggles

   A. when chipping stone
   B. when using a grinder
   C. while climbing or descending ladders
   D. when handling molten metal

7. When using a brace and bit to bore a hole completely through a partition, it is MOST important to

A. lean heavily on the brace and bit
B. maintain a steady turning speed all through the job
C. have the body in a position that will not be easily thrown off balance
D. reverse the direction of the bit at frequent intervals

8. Gloves should be used when handling

   A. lanterns   B. wooden rules
   C. heavy ropes   D. all small tools

Questions 9-16.

DIRECTIONS: Questions 9 through 16, inclusive, are based on the ladder safety rules given below. Read these rules fully before answering these items.

LADDER SAFETY RULES

When a ladder is placed on a slightly uneven supporting surface, use a flat piece of board or small wedge to even up the ladder feet. To secure the proper angle for resting a ladder, it should be placed so that the distance from the base of the ladder to the supporting wall is 1/4 the length of the ladder. To avoid overloading a ladder, only one person should work on a ladder at a time. Do not place a ladder in front of a door. When the top rung of a ladder rests against a pole, the ladder should be lashed securely. Clear loose stones or debris from the ground around the base of a ladder before climbing. While on a ladder, do not attempt to lean so that any part of the body, except arms or hands, extends more than 12 inches beyond the side rail. Always face the ladder when ascending or descending. When carrying ladders through buildings, watch for ceiling globes and lighting fixtures. Avoid the use of rolling ladders as scaffold supports.

9. A small wedge is used to

   A. even up the feet of a ladder resting on an uneven surface
   B. lock the wheels of a roller ladder
   C. secure the proper resting angle for a ladder
   D. secure a ladder against a pole

10. An 8 foot ladder resting against a wall should be so inclined that the distance between the base of the ladder and the wall is _____ feet.

    A. 2   B. 5   C. 7   D. 9

11. A ladder should be lashed securely when

    A. it is placed in front of a door
    B. loose stones are on the ground near the base of the ladder
    C. the top rung rests against a pole
    D. two people are working from the same ladder

12. Rolling ladders

    A. should be used for scaffold supports
    B. should not be used for scaffold supports
    C. are useful on uneven ground
    D. should be used against a pole

13. When carrying a ladder through a building, it is necessary to

    A. have two men to carry it
    B. carry the ladder vertically
    C. watch for ceiling globes
    D. face the ladder while carrying it

14. It is POOR practice to

    A. lash a ladder securely at any time
    B. clear debris from the base of a ladder before climbing
    C. even up the feet of a ladder resting on slightly uneven ground
    D. place a ladder in front of a door

15. A person on a ladder should NOT extend his head beyond the side rail by more than _____ inches.

    A. 12   B. 9   C. 7   D. 5

16. The MOST important reason for permitting only one person to work on a ladder at a time is that

    A. both could not face the ladder at one time
    B. the ladder will be overloaded
    C. time would be lost going up and down the ladder
    D. they would obstruct each other

17. Many portable electric power tools, such as electric drills, have a third conductor in the power lead which is used to connect the case of the tool to a grounded part of the electric outlet.
    The reason for this extra conductor is to

    A. have a spare wire in case one power wire should break
    B. strengthen the power lead so it cannot easily be damaged
    C. prevent the user of the tool from being shocked
    D. enable the tool to be used for long periods of time without overheating

18. Protective goggles should NOT be worn when

    A. standing on a ladder drilling a steel beam
    B. descending a ladder after completing a job
    C. chipping concrete near a third rail
    D. sharpening a cold chisel on a grinding stone

19. When the foot of an extension ladder, placed against a high wall, rests on a sidewalk or another such similar surface, it is advisable to tie a rope between the bottom rung of the ladder and a point on the wall opposite this rung.
    This is done to prevent

    A. people from walking under the ladder
    B. another worker from removing the ladder
    C. the ladder from vibrating when ascending or descending
    D. the foot of the ladder from slipping

4 (#2)

20. In construction work, practically all accidents can be blamed on the

    A. failure of an individual to give close attention to the job assigned to him
    B. use of improper tools
    C. lack of cooperation among the men in a gang
    D. fact that an incompetent man was placed in a key position

20.____

21. If it is necessary for you to do some work with your hands under a piece of heavy equipment while a fellow worker lifts up and holds one end of it by means of a pinch bar, one important precaution you should take is to

    A. wear gloves
    B. watch the bar to be ready if it slips
    C. insert a temporary block to support the piece
    D. work as fast as possible

21.____

22. Employees of the transit system whose work requires them to enter upon the tracks in the subway are cautioned not to wear loose fitting clothing.
    The MOST important reason for this caution is that loose fitting clothing may

    A. interfere when men are using heavy tools
    B. catch on some projection of a passing train
    C. tear more easily than snug fitting clothing
    D. give insufficient protection against subway dust

22.____

23. The MOST important reason for insisting on neatness in maintenance quarters is that it

    A. keeps the men busy in slack periods
    B. prevents tools from becoming rusty
    C. makes a good impression on visitors and officials
    D. decreases the chances of accidents to employees

23.____

24. Maintenance workers whose duties require them to do certain types of work generally work in pairs.
    The LEAST likely of the following possible reasons for this practice is that

    A. some of the work requires two men
    B. the men can help each other in case of accident
    C. there is too much equipment for one man to carry
    D. it protects against vandalism

24.____

25. A foreman reprimands a helper for actions in violation of the rules and regulations.
    The BEST reaction of the helper in this situation is to

    A. tell the foreman that he was careful and that he did not take any chances
    B. explain that he took this action to save time
    C. keep quiet and accept the criticism
    D. demand that the foreman show him the rule he violated

25.____

9

KEY (CORRECT ANSWERS)

1. A
2. D
3. D
4. D
5. A

6. C
7. C
8. C
9. A
10. A

11. C
12. B
13. C
14. D
15. A

16. B
17. C
18. B
19. D
20. A

21. C
22. B
23. D
24. D
25. C

SAFETY
EXAMINATION SECTION
TEST 1

DIRECTIONS: Each question or incomplete statement is followed by several suggested answers or completions. Select the one that BEST answers the question or completes the statement. *PRINT THE LETTER OF THE CORRECT ANSWER IN THE SPACE AT THE RIGHT.*

1. Following are three statements concerning safety in a hospital:
    I. Most accidents result from both unsafe actions of individuals and unsafe conditions
    II. The hospital security officer's major concern in accident investigation is the prevention of future accidents, not the fixing of blame
    III. A supervisor can determine his institution's safety performance by evaluating accident cost and frequency ratios against the same statistics from other comparable facilities

   Which one of the following CORRECTLY classifies the above statements into those which are correct and those which are not?

    A. I is correct, but II and III are not.
    B. III is correct, but I and II are not.
    C. I and II are correct, but III is not.
    D. I, II, and III are all correct.

2. Of the following types of portable hand-held fire extinguishers, the one which generally requires the extinguisher operator to stand closer to the fire because it has a shorter range than the others is

    A. carbon dioxide   B. foam
    C. soda-acid   D. dry chemical

3. Artificial respiration should be started IMMEDIATELY on a man who has suffered an electric shock if he is

    A. unconscious and breathing heavily
    B. unconscious and not breathing
    C. conscious and in a daze
    D. conscious and badly burned

4. According to a recent safety report, an outstanding cause of accidents to workers is the improper use of tools. The MOST helpful conclusion that you can draw from this statement is that

    A. most tools are dangerous to use
    B. many accidents from tools are unavoidable
    C. most tools are not used properly
    D. many accidents involving tools occur because of poor working habits

5. The MOST important reason employees are cautioned NOT to use water to put out fires in electrical equipment is that water

2 (#1)

    A. will cause corrosion of sensitive electrical parts
    B. will cause electrical shorts and result in fuses blowing
    C. may conduct the electrical current and create a shock hazard
    D. may generate dangerous gases when it comes into contact with hot insulation

6. When wood scaffolding is taken down, the nails are usually removed completely from the lumber.
The MAIN purpose of this procedure is to

    A. permit puttying up the nail holes
    B. prevent injury in future handling of the lumber
    C. avoid damage to the saw if the lumber is cut
    D. prevent the same old nails from being used on the next job

7. In an accident report, the information which may be MOST useful in decreasing the recurrence of similar type accidents is the

    A. extent of injuries sustained
    B. time the accident happened
    C. number of people involved
    D. cause of the accident

8. When putting out a fire with a hand extinguisher, it is BEST to direct the discharge at the _____ the fire.

    A. base of   B. area behind
    C. area in front of   D. highest flames of

9. Someone had suggested that the silver-plated main contacts of a circuit breaker be cleaned with fine sandpaper.
This suggestion is

    A. *poor,* since the useful silver plating would be removed
    B. *good,* since you would be removing silver oxide which is a poor conductor
    C. *good,* since this will prevent overheating of the circuit breaker
    D. *poor,* since this will change the adjustment of the main contacts

10. Of the following, the BEST way to lift a heavy object is to

    A. keep legs spread apart and straight, slowly bending at the waist to grasp the object
    B. place the feet about shoulder-width apart and slowly bend at the knees to reach down to the object
    C. keep legs straight and close together, slowly bending at the waist to grasp the object
    D. place feet close together, and with legs and back straight, bend at the waist to reach down and quickly lift the object

11. If you are assigned to work with someone who has recently caused an accident on the job, it is BEST for you to

    A. do most of the assigned work yourself
    B. personally check the safety precautions on the job
    C. refuse to work with the individual
    D. provide him with a copy of all rules and regulations

12. Sparks and open flames should be kept away from storage batteries that are being charged because of the high combustibility of the

    A. electrolytes in the batteries
    B. battery cases when hot
    C. gases being produced
    D. sulfuric acid fumes being generated

13. Of the following, the MOST important reason for requiring accident reports is that they will

    A. assist the attending physician
    B. assist in preventing future accidents
    C. teach employees to be systematic
    D. prevent lawsuits

14. Assume that a co-worker has collapsed in an area where monitoring equipment indicates that the air is poisonous. You remove the individual from the area and send a call for help. While waiting for help to arrive, you notice that the victim is not breathing adequately.
    Of the following, the FIRST action you should take is to

    A. raise the victim to a sitting position and give him strong coffee
    B. apply ice to the back of his neck
    C. keep him warm
    D. administer artificial respiration

15. Of the following types of fire extinguishers, the one that should be used to extinguish an electrical fire is the _____ fire extinguisher.

    A. soda-acid   B. foam
    C. carbon dioxide   D. water

16. The MAIN reason for grounding electrical equipment is to

    A. increase power to the coils
    B. increase wattage in the line
    C. prevent serious short circuits
    D. protect personnel from electrical shock

17. When disconnecting the electric wires from a motor, it is good practice to

    A. cut the live power wires
    B. assume that the circuit is alive
    C. scrape the terminals
    D. nick the wire in several places first

18. A safe procedure to follow when using a straight-type ladder is to

    A. use a box or stair to support the ladder legs
    B. hang tools on the ladder so they cannot be dropped
    C. take one step at a time when climbing
    D. have an assistant climb behind you to protect you

4 (#1)

19. Safety in work habits is MOST closely related to which of the following?   19._____

    A. Worker's knowledge of the job
    B. Speed with which a worker does the job
    C. Wages paid to a worker
    D. Carefulness of a worker

20. When a helper drops oil onto a plant floor, he should   20._____

    A. find out who is supposed to clean it up
    B. inform the foreman
    C. dry it up himself
    D. let it soak into the floor

21. Of the following materials, the BEST one to use as a tourniquet to stop bleeding from a severed artery is a(n)   21._____

    A. Venetian blind cord   B. electric extension cord
    C. leather belt   D. shoelace

22. Of the following types of portable fire extinguishers, the one that should NOT be used to extinguish a fire around a blower motor is the _____ extinguisher.   22._____

    A. dry chemical   B. carbon dioxide
    C. liquified gas   D. water solution

23. Workers whose characteristics and behavior are such as to make them considerably more liable to injury than the average person are considered by foremen to be   23._____

    A. late   B. safety conscious
    C. careful   D. accident-prone

24. Safety inspections by foremen are not useful in an accident prevention program unless   24._____

    A. all persons who have accidents are fined
    B. insurance rates are decreased
    C. immediate action is taken to correct the conditions revealed
    D. there is adequate compensation for all injured parties

25. If you replace a blown fuse, and the replaced fuse has burned out shortly thereafter, the FIRST step that should be taken when the replaced fuse has been damaged is that   25._____

    A. this second fuse should be replaced by a new fuse of the same type and amperage
    B. this second fuse should be replaced by a new fuse of slightly greater amperage
    C. the circuit should be disconnected while the cause of the burn-out is determined
    D. a check of all other fuses at the electrical connection should be made to determine if they were in working order

KEY (CORRECT ANSWERS)

1. D
2. A
3. B
4. D
5. C

6. B
7. D
8. A
9. A
10. B

11. B
12. C
13. B
14. D
15. C

16. D
17. B
18. C
19. D
20. C

21. C
22. D
23. D
24. C
25. C

TEST 2

DIRECTIONS: Each question or incomplete statement is followed by several suggested answers or completions. Select the one that BEST answers the question or completes the statement. *PRINT THE LETTER OF THE CORRECT ANSWER IN THE SPACE AT THE RIGHT.*

1. The BEST way for a supervisor to reduce the number of accidents among his workers is to

    A. draw up detailed safety procedures
    B. constantly inspect the operation for dangerous practices
    C. carefully investigate the cause of every accident
    D. develop safety consciousness in his workers

2. The MAIN reason for enclosing switches and other current-carrying parts of electrical systems is to

    A. exclude dirt and moisture
    B. protect workers from shock
    C. increase the capacity of the equipment
    D. prevent damage to the equipment

3. Employees who work near electrical equipment should avoid touching wires PARTICULARLY when wearing

    A. rubber-soled shoes
    B. wet clothing
    C. loose-fitting clothing
    D. nylon clothing

4. A leak in a chlorine tank is dangerous because chlorine

    A. may cause an explosion
    B. may corrode equipment
    C. is an irritating gas
    D. is highly flammable

5. Suppose that a member of your crew falls down a flight of stairs and severely injures his leg.
   While waiting for the doctor, you should try to

    A. make the injured man comfortable and prevent further injury
    B. put a splint on his leg if it appears to be broken
    C. question the injured man about the cause of the accident
    D. exercise the injured leg to prevent muscle tightening

6. Suppose that you see some acid splash into the eyes of a fellow worker.
   The one of the following actions that you should take FIRST is to

    A. bandage his eyes with a clean dressing
    B. place boric acid ointment in his eyes
    C. make him lie down and close his eyes
    D. wash his eyes with large amounts of clean water

7. Suppose that you discover an unconscious man who has fallen across a *live* electric wire.
Of the following, the action that you should take FIRST is to

   A. break the contact between the man and the live wire
   B. give the man artificial respiration
   C. loosen the man's clothing and treat him for shock
   D. apply sterile dressings to the man's burns

8. It is important to know the location of pressure points when treating persons suffering from

   A. shock   B. bleeding   C. poisoning   D. burns

9. The MAIN reason for treating sewage before emptying it into the ocean is to

   A. prevent pollution of the water
   B. remove disagreeable odors
   C. protect fishing areas
   D. obtain valuable by-products

10. The one of the following diseases which may be caused by the pollution of drinking water by sewage is

    A. malaria
    C. tuberculosis
    B. typhoid fever
    D. muscular dystrophy

11. When portable electrical equipment is grounded for safety purposes, the equipment ground wire usually has an outer covering colored

    A. black   B. red   C. yellow   D. green

12. Accident reports should be complete and detailed PRIMARILY to

    A. aid in fixing the blame where it belongs
    B. protect the supervisor's reputation
    C. furnish information that may be of help in accident prevention
    D. discipline subordinates who have been careless

13. Of the following, the MOST effective way to prevent violation of safety rules is to

    A. correct every infraction of a rule as soon as it is observed
    B. enforce each rule very strictly as soon as an accident has shown its importance
    C. dedicate a few days each month to strict enforcement
    D. avoid making an issue out of every infraction

14. Upon entering a pump room in which a motor-driven pump is running, the maintainer detects the odor of hot insulating varnish.
This odor indicates that the

    A. varnish has been freshly applied
    B. bearings are poorly lubricated
    C. room is insufficiently ventilated
    D. motor is being overloaded

15. Maintenance workers are required to report defective equipment to their superiors even when the maintenance of the particular equipment is handled by another bureau.
    The purpose of this rule is to

    A. reward those who keep their eyes open
    B. punish employees who don't do their jobs
    C. have repairs made before serious trouble occurs
    D. keep employees on their toes

16. The MAIN reason for not permitting more than one person to work on a ladder at the same time is that

    A. the ladder might get overloaded
    B. several persons on the ladder might obstruct each other
    C. time would be lost going up and down the ladder
    D. several persons could not all face the ladder at one time

17. Safety on the job is BEST assured by

    A. keeping alert
    B. working only with new tools
    C. working very slowly
    D. avoiding the necessity for working overtime

18. A serious safety hazard occurs when a _____ hammer is used to strike a _____ surface.

    A. hardened steel; hardened steel
    B. soft iron; hardened steel
    C. hardened steel; soft iron
    D. soft iron; soft iron

19. When an emergency exit door set in the sidewalk is being opened from inside, the door should be opened *slowly* to avoid

    A. injury to pedestrians
    B. making unnecessary noise
    C. a sudden rush of air from the street
    D. damage to the sidewalk

20. The MAIN purpose of the periodic inspections of facilities and equipment that are made by maintainers is probably to

    A. encourage the men to take better care of these facilities and equipment
    B. keep the maintainers busy during otherwise slack periods
    C. discover minor faults before they develop into more serious conditions
    D. make the men more familiar with these facilities and equipment

21. After pulling the fuse of a 600-volt circuit, and before starting the work of connecting additional equipment to the circuit, the MOST important safety precaution to take is to

    A. examine the condition of the fuse
    B. disconnect all load from the circuit
    C. check that all tools have insulated handles
    D. test to make sure the circuit is dead

22. The MOST practical way to determine in the field if a large coil of #14 wire has the required length for a given job is to

   A. weigh the coil and compare with a new 1000-foot coil
   B. measure the electrical resistance and compare with a 1000-foot coil
   C. measure the length of one turn and multiply by the number of turns
   D. unwind the coil and lay the wire alongside the conduit before pulling it in

22.____

23. Maintainers of the transit system are required to report defective equipment to their superiors, even when the maintenance of the particular equipment is handled entirely by another bureau.
The purpose of this rule is to

   A. fix responsibility
   B. discourage slackers
   C. encourage alertness
   D. prevent accidents

23.____

24. The MOST important reason for covering a wood door with sheet metal is to make the door more

   A. burglar-proof
   B. fire-resistant
   C. termite-proof
   D. resistant to natural decay and deterioration

24.____

25. An employee will MOST likely avoid accidental injury if he

   A. stops to rest frequently
   B. works alone
   C. keeps mentally alert
   D. works very slowly

25.____

KEY (CORRECT ANSWERS)

1.	D	11.	D
2.	B	12.	C
3.	B	13.	A
4.	C	14.	D
5.	A	15.	C
6.	D	16.	A
7.	A	17.	A
8.	B	18.	A
9.	A	19.	A
10.	B	20.	C

21. D
22. C
23. D
24. B
25. C

EXAMINATION SECTION
TEST 1

DIRECTIONS: Each question or incomplete statement is followed by several suggested answers or completions. Select the one that BEST answers the question or completes the statement. *PRINT THE LETTER OF THE CORRECT ANSWER IN THE SPACE AT THE RIGHT.*

1. Of the following, the MOST important objective in accident analysis is to determine

    A. who is to blame
    B. how the accident could have been prevented
    C. whether the injured persons had received prescribed medical treatment
    D. the names of the persons involved and the exact nature of the injuries

2. Of the following statements, the one which would be of LEAST help in getting your men to do their work safely is to

    A. correct their unsafe work habits only if you think they may cause an accident
    B. see that they read and understand the safety rules they should follow
    C. talk to them often about the important of following safety regulations
    D. watch them at all times when they are working to see that they observe safety rules

3. The safety experts in your agency want to study accident reports as they are submitted in order to learn ways of preventing future accidents. You have been assigned to design a new accident report form which will help them achieve this goal.
Of the following, the item in the report that would be of MOST value to your safety experts for their purpose is

    A. the name and age of the accident victim
    B. a description of the working situation which led to the accident
    C. a statement from the foreman whether there was any carelessness involved
    D. a statement by the employee involved as to how the accident might have been prevented

4. A foreman should investigate every accident, including minor ones that do not involve injury, MAINLY because

    A. investigation of all accidents will help to increase the safety awareness of the employees
    B. each accident indicates a potential source of injury or damage
    C. the foreman will receive valuable experience in spotting hazardous conditions
    D. safety records should include all accidents regardless of seriousness

5. The following are methods used to prevent accidents. The one that is the MOST effective way to provide for safe and efficient operation of machines is to

    A. station guards on machines where physical hazards exist
    B. train employees in job procedures that will minimize accidents
    C. provide protective clothing and equipment for work on dangerous machines
    D. eliminate hazards by including safety considerations in the basic design

6. The MOST important element of preventive maintenance is

   A. calibration
   B. lubrication
   C. inspection
   D. cleaning

7. Of the following statements, the one that is MOST accurate is that a supervisor is

   A. not responsible for power tools used by his men
   B. not responsible for power tools used by his men if the men are properly trained in the use of these power tools
   C. always responsible for power tools used by his men
   D. responsible for power tools used by his men only if there has not been adequate time to train the men in the use of the power tools

8. Assume that your crew has been issued an item of safety equipment and the men refuse to wear it because it is not the brand they are used to.
   You, as supervisor, should

   A. let them work without it until you check with your superior
   B. stop the men from working and report the facts to your superior immediately
   C. warn them that they may lose compensation payments if there is an accident
   D. have the supply man issue the proper equipment

9. A steel measuring tape is undesirable for use around electrical equipment.
   The LEAST important reason is the

   A. magnetic effect
   B. short circuit hazard
   C. shock hazard
   D. danger of entanglement in rotating machines

10. When using a portable extension cord, the MOST important precaution to take is to

    A. see that the cord does not create a tripping hazard
    B. make sure that the cord does not touch any metal
    C. have a polarized plug at the end of the cord
    D. keep the cord clean and dry

11. Of the following, the one which is MOST likely to have the GREATEST effect in improving safety is

    A. holding foremen accountable for accidents of subordinates
    B. periodic safety inspections
    C. posting numerous safety bulletins
    D. providing each worker with periodic safety newsletters

12. The proper extinguisher to use on an electrical fire in an operating electric motor is

    A. sand
    B. water
    C. soda and acid
    D. carbon dioxide

13. When a container is used for flammable liquids, it usually presents the GREATEST hazard when it is

    A. empty but uncleaned
    B. empty and clean
    C. filled
    D. half-filled

14. Of the following cans, the SAFEST type of can to use for storing oil-soaked rags indoors is

    A. perforated sheet metal can with a sheet metal cover
    B. sheet metal can with a sheet metal cover
    C. sheet metal can without a cover
    D. sheet metal can with perforated sheet metal cover

15. To take the strain off the connections to an electrical plug, the knot which should be used is a(n)

    A. bowline
    B. hitch
    C. underwriter's
    D. square

16. In analyzing safety performance, the term *injury frequency rate* is used. This is defined as the number of disabling injuries per 1,000,000 man-hours worked.
    If an assistant supervisor has 80 men under him working 40 hours a week and 5 men suffered disabling injuries in a working period of 52 weeks, then the injury frequency rate would be closest to

    A. 25   B. 30   C. 35   D. 40

17. Of the following, the organization that MOST often certifies to the safety of individual pieces of equipment is the

    A. American Society for Testing Materials
    B. Underwriters' Laboratories, Inc.
    C. American Standards Association
    D. Association of Casualty and Surety Companies

18. The extinguishing agent in a soda-acid fire extinguisher is

    A. water
    B. carbon dioxide
    C. carbon tetrachloride
    D. calcium chloride solution

19. The proper technique for lifting heavy objects includes all of the following EXCEPT

    A. bending the knees
    B. placing the feet as far from the object as possible
    C. keeping the back straight
    D. lifting with the arras and legs

20. The one of the following methods which should NOT be used in treating portable wooden ladders is

    A. the application of a coat of clear lacquer
    B. thorough washing with soap and water
    C. the application of a coat of white paint
    D. the application of a coat of linseed oil

21. The MAIN reason for a requirement that defective material be removed from a job site as soon as possible is to   21.____

    A. prevent injuries
    B. reduce clutter in the area
    C. prevent accidental use of the material
    D. permit more efficient operation

22. The MAIN reason for reporting accidents is to   22.____

    A. prevent future accidents of the same type
    B. determine who was at fault
    C. prevent unwarranted lawsuits
    D. have a record of the causes of delays

23. Employees who must lift and carry stock items should be careful to avoid injury. When an employee lifts or carries stock items, which of the following is the LEAST safe practice?   23.____

    A. Keep the legs straight and lift with the back muscles
    B. Keep the load as close to the body as possible
    C. Get a good grip on the object to be carried
    D. First determine if the item can be lifted and carried safely

24. For warning and protection, the color *red* is usually for   24.____

    A. indicating high temperature stockroom areas
    B. floor markings
    C. location of first-aid supplies
    D. stop buttons, lights for barricades, and other dangerous locations

25. Reporting rattles, squeaks, or other noises in equipment to your maintenance supervisor is   25.____

    A. *bad;* too much attention to squeaks like these keep important safety problems from being noticed
    B. *bad;* each person should oil and care for his own equipment
    C. *good;* these sounds may mean that the equipment should be fixed
    D. *good;* it shows the supervisor that you are a good worker

KEY (CORRECT ANSWERS)

1. B
2. A
3. B
4. B
5. D

6. C
7. C
8. B
9. A
10. A

11. B
12. D
13. A
14. B
15. C

16. B
17. B
18. A
19. B
20. C

21. C
22. A
23. A
24. D
25. C

TEST 2

DIRECTIONS: Each question or incomplete statement is followed by several suggested answers or completions. Select the one that BEST answers the question or completes the statement. *PRINT THE LETTER OF THE CORRECT ANSWER IN THE SPACE AT THE RIGHT.*

1. An agency gives some of its maintenance employees instruction in first aid. The MOST likely reason for doing this is to

    A. eliminate the need for calling a doctor in case of accident
    B. reduce the number of accidents
    C. lower the cost of accidents to the agency
    D. provide temporary first aid

    1.____

2. If a fellow worker has stopped breathing after an electric shock, the BEST first-aid treatment is

    A. artificial respiration
    B. to massage his chest
    C. an application of cold compresses
    D. a hot drink

    2.____

3. If you had to telephone for an ambulance because of an accident, the MOST important information for you to give the person who answered the telephone would be the

    A. exact time of the accident
    B. place where the ambulance is needed
    C. cause of the accident
    D. names and addresses of those injured

    3.____

4. To use clean ice water as a treatment for burns is

    A. *good*, because it gives immediate relief from pain and seems to lessen the damaging effects of burns
    B. *bad*, because it has a tendency to cause frostbite which may develop into gangrene
    C. *good*, because ice water will destroy any bacteria at once
    D. *bad*, because the extremely cold temperature will cause a person to go into shock

    4.____

5. One of your men strained the muscles in his back when he attempted to lift a load that was extremely heavy.
   The treatment for this injury would include all of the following EXCEPT

    A. applying cold cloths or an ice bag to the back
    B. massaging the area
    C. resting the back in its most comfortable position
    D. rubbing the strained muscles with witch hazel

    5.____

6. A man who fainted in the terminal is now semiconscious. He is bleeding about the mouth and is in danger of choking on the blood.
   He should be placed on his

    6.____

2 (#2)

    A. back, with his head slightly lower than his feet
    B. stomach, with his head turned to one side, lower than his feet
    C. back, with his feet slightly lower than his head
    D. stomach, with his head turned to one side, higher than his feet

7. When administering first aid to a helper suffering from shock as a result of falling off a high ladder, it is MOST important to

    A. cover the helper and keep him warm
    B. give the helper something to drink
    C. apply artificial respiration to the helper
    D. prop the helper up to a sitting position

8. If a co-worker's clothing gets caught in the gears of a machine in operation, the FIRST thing for a helper to do is to

    A. call the supervisor
    B. try to pull him out
    C. shut off the machine's power
    D. jam a metal tool between the gears of the machine

9. The one of the following which is the FIRST thing to do when a person gets an electric shock and is still in contact with the supply is to

    A. start artificial respiration immediately
    B. treat for burns
    C. cut the power if it takes no more than 5 minutes to locate the switch
    D. remove the victim from the contact by using a dry stick or dry rope

10. The one of the following that is the LEAST important health precaution for a worker to take is

    A. frequent washing
    B. shading his eyes from reflected light
    C. using an antiseptic on cuts
    D. wearing rubber gloves

11. Before entering a sewer which is known to contain dangerous gases, the able foreman will

    A. drop lighted matches down the manhole
    B. make sure all manholes in the vicinity are closed
    C. send one man down to determine the amount of gas present
    D. wait until the sewer has been ventilated

12. The one of the following that would MOST likely be the cause of a sewer explosion is

    A. a pressure relief valve installed in the main sewer line
    B. an unplugged opening left for a house connection
    C. naphtha discharged into the sewer by a cleaning establishment
    D. sewage of recent origin containing dissolved oxygen

13. In case of severe injury and where there is a possibility of broken bones, the MOST important precaution to take in giving first aid to an injured man is:

    A. Bundle him into an automobile and get him to a hospital as fast as possible
    B. Lower his feet and raise his head
    C. Move him no more than necessary and call a doctor
    D. Raise him to a sitting position and give him a drink of water

14. The logical reason that certain employees who work on the tracks carry small parts in fiber pails rather than steel pails is that fiber pails

    A. are stronger
    B. can't rust
    C. can't be dented by rough usage
    D. do not conduct electricity

15. While working on a certain track between stations, a helper notices a man standing on an adjacent track and suspects from the man's actions that he may have no business being there.
    The MOST reasonable procedure would be to

    A. continue working and ignore the man
    B. order the man to get off the tracks immediately
    C. ask the man what business he has being there
    D. hold the man for questioning by police

16. With respect to safety of personnel, it is probably LEAST important to

    A. have a place for each tool and put each tool in its place at the end of each day
    B. place each tool where it cannot fall down and hurt anyone when working on a job
    C. coat each tool with grease at the end of each day to prevent rust
    D. inspect carefully all tools to be used before beginning the day's work

17. Employees whose work requires them to enter upon the tracks are cautioned not to wear loose-fitting clothing. The MOST important reason for this caution is that loose-fitting clothing may

    A. interfere when they are using heavy tools
    B. catch on some projection of a passing train
    C. give insufficient protection against dust
    D. tear more easily than snug-fitting clothing

18. Recent safety reports indicate that a principal cause of injury to employees is *falls* while on a job. Such reports tend to emphasize that safety on the job is BEST assured by

    A. following every rule   B. keeping alert
    C. never working alone   D. working very slowly

19. The one of the following statements about a plug fuse that is MOST valid is that it should

    A. always be screwed in lightly to assure easy removal
    B. never be used to hold a coin in the fuse socket
    C. never be replaced by someone unfamiliar with the circuit
    D. always be replaced by a larger size if it burns out frequently

20. If a helper has frequent accidents, it is MOST likely that he is

   A. not physically strong enough to do the job
   B. simply one of those persons who is unlucky
   C. not paying enough attention to safe work habits
   D. trying too hard

21. A rule states that, *In walking on the track, walk opposite to the direction of traffic on that track if possible.*
   By logical reasoning, the PRINCIPAL safety idea behind this rule is that the man on the track

   A. is more likely to see an approaching train
   B. will be seen more readily by the motorman
   C. need not be as careful
   D. is better able to judge the speed of the train

22. The PRINCIPAL objection to using water from a hose to put out a fire involving live electrical equipment is that

   A. insulation may be damaged
   B. cast iron parts may rust
   C. serious electric shock may result
   D. a short-circuit will result

23. An electrician's knife should NOT be used to

   A. cut copper wires
   B. remove rubber insulation
   C. cut friction tape
   D. sharpen pencils

24. According to a safety report, a frequent cause of accidents to workers is the improper use of tools.
   The MOST helpful conclusion that you can draw from this statement is that

   A. most tools are difficult to use properly
   B. most tools are dangerous to use
   C. many accidents from tools are unavoidable
   D. many accidents from tools occur because of poor working habits

25. When a maintainer reports a minor trouble orally to his foreman, the MOST important information the foreman would require from the maintainer would be the

   A. type of trouble and its exact location
   B. names of all men with him when he discovered the trouble
   C. exact time the trouble was discovered
   D. work he was doing when he noted the trouble

KEY (CORRECT ANSWERS)

1.	D		11.	D
2.	A		12.	C
3.	B		13.	C
4.	A		14.	D
5.	A		15.	C
6.	B		16.	C
7.	A		17.	B
8.	C		18.	B
9.	D		19.	B
10.	B		20.	C

21. A
22. C
23. A
24. D
25. A

SAFETY
EXAMINATION SECTION
TEST 1

DIRECTIONS: Each question or incomplete statement is followed by several suggested answers or completions. Select the one that BEST answers the question or completes the statement. *PRINT THE LETTER OF THE CORRECT ANSWER IN THE SPACE AT THE RIGHT.*

1. There are two indicators used to determine the safety record of an agency. One is the "frequency of injury," and the other is the "severity of injury."
   The "frequency of injury" is considered a better indicator of the safety record because

   A. blind chance has a greater effect on "severity" than on "frequency"
   B. it is easier to record "frequency" than "severity"
   C. workers will pay more attention to "frequency" than to "severity"
   D. it is more difficult to determine the "severity" than the "frequency"

   1._____

2. It is frequently said that some people are "accident prone." This term should be applied ONLY to those people who

   A. fail to respond to safety training
   B. have accidents when the cause of the accident cannot be determined
   C. lack the physical capacity for their job
   D. do not have the skill required to do a certain job

   2._____

3. "Accidents frequently happen because a man *daydreams* on the job." Of the following, the one that is CORRECT based on the previous sentence is:

   A. Accidents are most often caused by *daydreaming*
   B. The main cause of poor work is accidents
   C. A man who does not *daydream* is a good worker
   D. It is important for a man to pay attention to what he is doing

   3._____

4. Accidents can be classified as caused either by "unsafe acts" or "unsafe conditions." The one of the following that would be considered as "unsafe condition" is

   A. jumping over an obstruction on the floor
   B. poor lighting in a crowded cellar
   C. speeding in a motor vehicle
   D. use of the wrong tool for a job

   4._____

5. Of the following types of fires, a soda-acid fire extinguisher is NOT recommended for

   A. electric motor controls   B. waste paper
   C. waste rags   D. wood desks

   5._____

6. A foam-type fire extinguisher extinguishes fires by

   A. cooling only   B. drenching only
   C. smothering only   D. cooling and smothering

   6._____

31

7. If an air-conditioning unit shorted out and caught fire, the BEST fire extinguisher to use would be _____ extinguisher.

   A. water
   B. foam
   C. carbon dioxide
   D. soda acid

8. The one of the following diseases which may be caused by the pollution of drinking water by sewage is

   A. malaria
   B. typhoid fever
   C. tuberculosis
   D. muscular dystrophy

9. The type of portable fire extinguisher that is MOST effective in controlling a fire around live electrical equipment is the

   A. foam type
   B. soda-acid type
   C. carbon-dioxide type
   D. water type

10. The hazards of electric shock resulting from operation of a portable electric tool in a damp location can be *reduced* by

    A. grounding the tool
    B. holding the tool with one hand
    C. running the tool at low speed
    D. using a baffle

11. The MAIN reason caretakers are advised to always wear protective goggles while changing a broken bulb is to avoid the danger of

    A. glare from the bulb
    B. pieces of glass getting in the eyes
    C. sparks from the bulb
    D. insects on or around the bulb socket

12. Of the following types of fire extinguishers, the one to use on an electrical fire is

    A. soda acid
    B. carbon dioxide
    C. water pump tank
    D. pyrene

13. The GREATEST number of injuries from equipment used in construction work result from

    A. carelessness of the operator
    B. poor maintenance of the equipment
    C. overloading of the equipment
    D. poor inspection of the equipment

14. Of the following, the BEST way a laborer can avoid accidents is to

    A. work slowly
    B. be alert
    C. wear safety shoes
    D. wear glasses

15. Of the following actions, the BEST one to take FIRST after smoke is seen coming from an electric control device is to

   A. shut off the power to it
   B. call the main office for advice
   C. look for a wiring diagram
   D. throw water on it

16. Of the following fire extinguishers, the one which should be provided for use in the elevator machine room is the

   A. carbon-dioxide type
   B. soda-acid type
   C. foam type
   D. loaded-stream type

17. Frequent deaths are reported as a result of running an automobile engine in a closed garage. Death results from

   A. suffocation
   B. carbon monoxide poisoning
   C. excessive humidity
   D. an excess of carbon dioxide in the air

18. As a veteran sewage treatment worker, you can BEST promote safety in your operations by

   A. carefully investigating and reporting the circumstances of any accident
   B. suggesting safer methods of operation
   C. training subordinates in proper safety
   D. disciplining subordinates who engage in unsafe acts

19. Oil soaked rags are BEST stored in a

   A. neat pile in a readily accessible corner
   B. metal container with a tight cover
   C. metal box that has holes for adequate ventilation
   D. closet on a shelf above the ground

20. The one of the following actions that is NOT the cause of injury when working with hand tools is

   A. working with defective tools
   B. using the wrong tool for the job
   C. working too carefully
   D. using a tool improperly

21. To safely lift a heavy object from the ground, you should keep your arms and elbows

   A. away from the body with your back bent
   B. away from the body with your back straight
   C. close to the body with your back bent
   D. close to the body with your back straight

Questions 22-25.

DIRECTIONS: Each question consists of a statement. You are to indicate whether the statement is TRUE (T) or FALSE (F). *PRINT THE LETTER OF THE CORRECT ANSWER IN THE SPACE AT THE RIGHT.*

22. The foam type extinguisher is not suitable for use on gasoline fires.   22.____

23. The first thing an employee should do when he sees a smoking electric wire is to throw water on the wire.   23.____

24. Many accidents are caused by carelessness of employees while at work.   24.____

25. If, at work, you are unable to lift a very heavy object, you should rest a couple of minutes and try again.   25.____

KEY (CORRECT ANSWERS)

1.	A	11.	B
2.	A	12.	B
3.	D	13.	A
4.	B	14.	B
5.	A	15.	A
6.	D	16.	A
7.	C	17.	B
8.	B	18.	C
9.	C	19.	B
10.	A	20.	C

21. D
22. F
23. F
24. T
25. F

TEST 2

DIRECTIONS: Each question or incomplete statement is followed by several suggested answers or completions. Select the one that BEST answers the question or completes the statement. *PRINT THE LETTER OF THE CORRECT ANSWER IN THE SPACE AT THE RIGHT.*

1. Assume that a fire breaks out in an electrical control panel board. Of the following types of portable fire extinguishers, the BEST one to use to put out this fire would be a

    A. dry-chemical type
    B. soda-acid type
    C. foam type
    D. water-stream type

2. The MAJORITY of home accidents result from

    A. burns
    B. suffocation
    C. falls
    D. poisons

3. A soda-acid fire extinguisher is recommended for use on fires consisting of

    A. wood or paper
    B. fuel oil or gasoline
    C. electrical causes or fuel oil
    D. paint or turpentine

4. Of the following, the extinguishing agent that should be used on fires in flammable liquids is

    A. steam
    B. water
    C. foam
    D. soda and acid

5. Of the following, the BEST way to put out a gasoline fire is to use

    A. a carbon dioxide extinguisher
    B. compressed air
    C. water
    D. rags to smother the blaze

6. A heavy object should be lifted by first crouching and firmly grasping the object to be lifted. Then, the worker should lift

    A. using his back muscles and keeping his legs bent
    B. by straightening his legs and keeping his back as straight as possible
    C. using his arm muscles and keeping his back nearly horizontal
    D. using his arm muscles and keeping his feet close together

7. The proper type of firefighting equipment to be used on an electrical fire is a

    A. soda-acid type extinguisher
    B. fire hose and water
    C. dry-chemical type extinguisher
    D. foam type extinguisher

8. While working on the job, you accidentally break a window pane. No one is around, and you are able to clean up the broken pieces of glass. It would then be BEST for you to

   A. leave a note near the window that a new glass has to be put in because it was accidentally broken
   B. forget about the whole thing because the window was not broken on purpose
   C. write a report to your supervisor telling him that you saw a broken window pane that has to be fixed
   D. tell your supervisor that you accidentally broke the window pane while working

8.____

9. The BEST way to remove some small pieces of broken glass from a floor is to

   A. use a brush and dust pan
   B. pick up the pieces carefully with your hands
   C. use a wet mop and a wringer
   D. sweep the pieces into the corner of the room

9.____

10. Employees should wipe up water spilled on floors immediately. The BEST reason for this is that water on a floor

    A. is a sign that employees are sloppy
    B. makes for a slippery condition that could cause an accident
    C. will eat into the wax protecting the floor
    D. is against health regulations

10.____

11. A carbon dioxide fire extinguisher is BEST suited for extinguishing

    A. paper fires   B. rag fires
    C. rubbish fires   D. grease fires

11.____

12. A pressurized water or soda-acid fire extinguisher is BEST suited for extinguishing

    A. wood fires   B. gasoline fires
    C. electrical fires   D. magnesium fires

12.____

13. Assume that an officer, alone in a building at night, smells the strong odor of cooking or heating gas. In addition to airing the building and making sure that he is not overcome, it would be BEST for the officer to call

    A. his superior at his home and ask for instructions
    B. for a plumber from the department of public works
    C. 911 for police and fire help
    D. the emergency number at Con Edison

13.____

14. The one of the following which is the MOST common safety hazard in an office is

    A. a sharp pencil on a desk
    B. an open desk drawer
    C. lack of covers for electric computers
    D. lack of parallel alignment of desks

14.____

15. Which of the following situations is MOST likely to pose the greatest danger to safety?

    A. Buffing a main corridor to a high shine
    B. Leaving a door to a hall open at a 180 angle

15.____

36

3 (#2)

C. Opening the top two drawers of a four-drawer file cabinet
D. Setting a desk at a 45 angle near a main aisle in an office

16. Safety experts agree that accidents can probably BEST be prevented by   16._____

   A. developing safety consciousness among employees
   B. developing a program which publicizes major accidents
   C. penalizing employees the first time they do not follow safety procedures
   D. giving recognition to employees with accident-free records

17. The accident records of many agencies indicate that most on-the-job injuries are caused   17._____
    by the unsafe acts of their employees. Which one of the following statements pinpoints
    the *most probable* cause of this safety problem?

   A. Responsibility for preventing on-the-job accidents has not been delegated.
   B. Lack of proper supervision has permitted these unsafe actions to continue.
   C. No consideration has been given to eliminating environmental job hazards.
   D. Penalties for causing on-the-job accidents are not sufficiently severe.

18. Which of the following methods is LEAST essential to the success of an accident prevention program?   18._____

   A. Determining corrective measures by analyzing the causes of accidents and making recommendations to eliminate them
   B. Educating employees as to the importance of safe working conditions and methods
   C. Determining accident causes by seeking out the conditions from which each accident has developed
   D. Holding each supervisor responsible for accidents occurring during the on-the-job performance of his immediate subordinates

19. Assume that you have a bad cold and take a strong decongestant pill before you come to   19._____
    work. You are scheduled, that day, to drive an official car to a supermarket to make an
    inspection. Of the following, it would be BEST for you to

   A. drive to the store and make the inspection as usual
   B. drive to the store very slowly and carefully, since you are not feeling well
   C. explain to your supervisor that you should not drive that day
   D. start out to make the inspection, but return to the office if you feel your driving ability is impaired

20. Of the following office supplies, the kind which you should usually be MOST careful to   20._____
    keep away from an open flame is

   A. carbon paper   B. ink
   C. paste   D. typing paper

21. The only one of the following types of fire extinguishers which should generally NOT be   21._____
    used to extinguish a gasoline fire is

   A. carbon dioxide   B. dry chemical
   C. foam   D. water

Questions 22-25.

DIRECTIONS: Each question consists of a statement. You are to indicate whether the statement is TRUE (T) or FALSE (F). *PRINT THE LETTER OF THE CORRECT ANSWER IN THE SPACE AT THE RIGHT.*

22. In directing the stream from a foam-type extinguisher at a fire, the extinguisher should be held upside down.   22.____

23. The carbon tetrachloride type of extinguisher is the most effective for use on electrical fires.   23.____

24. A soda-and-acid type of extinguisher should be refilled at least once in five years.   24.____

25. In lifting heavy articles, sanitation men should keep their feet wide apart.   25.____

KEY (CORRECT ANSWERS)

1.	A	11.	D
2.	C	12.	A
3.	A	13.	D
4.	C	14.	B
5.	A	15.	C
6.	B	16.	A
7.	C	17.	B
8.	D	18.	D
9.	A	19.	C
10.	B	20.	A

21. D
22. T
23. T
24. F
25. F

SAFETY
EXAMINATION SECTION
TEST 1

DIRECTIONS: Each question or incomplete statement is followed by several suggested answers or completions. Select the one that BEST answers the question or completes the statement. *PRINT THE LETTER OF THE CORRECT ANSWER IN THE SPACE AT THE RIGHT.*

1. In lifting a heavy can, a caretaker should keep his

   A. back and knees straight
   B. back bent and knees straight
   C. knees and back bent
   D. knees bent and back straight

1.____

2. If a man is injured on the job and it is likely that he has broken bones, the foreman should

   A. call for an ambulance
   B. call the superintendent
   C. take him to the hospital in his car
   D. tell the injured man to go to the hospital immediately

2.____

3. The MAIN reason for not letting dust cloths or oily rags pile up in storage closets is that

   A. a fire may start from this material
   B. the closet will not look neat and orderly
   C. the oil may soak into the floor and stain it
   D. they take up valuable space which may be put to better uses

3.____

4. Suppose, in making your rounds, you come upon a small oil and grease fire in a basement. After putting in a fire alarm, you find the fire extinguisher is out of order. The BEST thing for you to do is to

   A. do nothing since you have put in an alarm
   B. open all the doors and windows
   C. throw earth and sand on the fire
   D. throw water on the fire

4.____

5. The BEST thing to do for a man who feels he is about to faint is to

   A. apply a cold compress to his forehead
   B. give him some cold water to drink
   C. lower his head between his knees
   D. move him out to the fresh air

5.____

6. In removing the end of a broken bulb from a socket, the caretaker should stick a hard rubber wedge into the socket and

   A. pull the wedge down
   B. push the wedge up

6.____

39

C. turn the wedge to the right
D. turn the wedge to the left

7. Employees are required to wear gloves when handling metal plates. The MAIN purpose for this is to minimize the possibility of

   A. smudging the metal plates
   B. dropping the metal plates
   C. strained wrist muscles
   D. hand cuts caused by the metal

8. A report of a personal injury written by the injured employee is MOST likely to be factually correct if written

   A. after getting the versions of witnesses
   B. upon the return of the employee from the doctor
   C. after discussing it with the foreman
   D. immediately after the injury

9. An employee notices an unsafe condition in his/her work area. Which of the following should he/she do FIRST?

   A. Post a sign warning of the condition.
   B. Report the condition to his foreman.
   C. Tell his fellow workers about the condition.
   D. Rope off the area.

10. You notice an oil spill in your aisle space adjacent to your work area. Of the following, which is the BEST action for you to take?

    A. Mop the spill up with water.
    B. Mark the area with a white warning cloth.
    C. Rope off the area.
    D. Cover the oil spill with Speedi-dry or Zorball.

11. A keg of nails is too heavy for you to lift alone. Of the following, which is the BEST action for you to take?

    A. Ask another employee to help you lift it.
    B. Bend your knees before lifting it.
    C. Ask your foreman to assign a stronger employee to lift it.
    D. Open the keg and subdivide the contents into smaller quantities.

12. A subordinate sustains a deep gash on his arm. Which of the following is the BEST action for you to take?

    A. Give him first aid, then instruct him to continue working.
    B. Give him first aid, then send him to a clinic.
    C. Tell him to get the first aid kit and treat himself.
    D. Tell him to go to his own doctor.

13. The PRIMARY reason for keeping storeroom aisles clear of obstructions is to 13._____

    A. enhance inventory control
    B. have a clear view of the entire stockroom
    C. minimize accidents
    D. make it easier to clear the aisles

14. Flammable material should be issued in what type of container? 14._____
    A(n) _____ container.

    A. glass   B. metal
    C. approved safety   D. plastic

15. While pouring a chemical fluid into a pail of water, a cleaner splashes some of the fluid in 15._____
    his eye.
    The FIRST thing he should do is wash his eye with

    A. a salve   B. a light oil
    C. rubbing alcohol   D. running water

16. A cleaner threw an oily rag into the corner of the cleaner's room, intending to leave it 16._____
    there until the next time he needed it.
    His action was

    A. *proper* because he could easily find it when he wanted it
    B. *improper* because it made the room look untidy
    C. *proper* because it was out of the sight of passengers
    D. *improper* because oily rags can easily catch fire

17. If a cleaner notices that any light bulbs are burned out, he is required to report it immedi- 17._____
    ately.
    The MAIN reason for promptly reporting this condition is to

    A. avoid complaints
    B. keep his supervisor informed
    C. prevent accidents
    D. make a good impression on his supervisor

18. Cleaners must be very careful not to let water get into escalator pits and on escalator 18._____
    machinery, or on the electrical components of escalators.
    The MAIN reason for this precaution is that

    A. water may cause rust
    B. electrical parts do not need cleaning
    C. water and oil do not mix
    D. water may cause a short circuit

19. The MAIN reason for keeping a storage cabinet locked at all times is to keep out 19._____

    A. noise   B. drafts of air
    C. other employees   D. thieves

20. In stopping a motor vehicle on snow-covered streets, the OBJECTION to disengaging the clutch and applying the brakes hard at the same time is that 20.____

    A. the abuse of tires under such conditions is greater
    B. it causes too great a loss of power
    C. it permits the motor to race which causes overheating
    D. the possibility of skidding is increased

KEY (CORRECT ANSWERS)

1.	D	11.	A
2.	A	12.	B
3.	A	13.	C
4.	C	14.	C
5.	C	15.	D
6.	D	16.	D
7.	D	17.	C
8.	D	18.	D
9.	B	19.	D
10.	D	20.	D

TEST 2

DIRECTIONS: Each question or incomplete statement is followed by several suggested answers or completions. Select the one that BEST answers the question or completes the statement. *PRINT THE LETTER OF THE CORRECT ANSWER IN THE SPACE AT THE RIGHT.*

1. Of the following potential fire sources, the one which would MOST likely produce a Class D fire is

    A. a flammable liquid
    B. faulty electrical wiring
    C. dry rubber
    D. a combustible metal

    1.____

2. A LIKELY result of using a carbon dioxide extinguishing agent to put out an electrical fire would be

    A. extensive weakening of the flooring
    B. successful extinguishment of the fire
    C. blown fuses or tripped circuit breakers
    D. conduction of electricity creating more potential danger

    2.____

3. A common symptom of shock visible in an accident victim's eyes is dilated pupils.
   The term *dilated,* as used in the above sentence, means

    A. enlarged
    B. insensitive
    C. constricted
    D. lackluster

    3.____

4. When using the mouth-to-mouth technique of artificial respiration, the person administering first aid should blow air into the victim's lungs at an average rate of _____ breaths per minute for an adult victim and _____ breaths per minute for a child.

    A. 12; 20   B. 20; 10   C. 20; 30   D. 30; 60

    4.____

5. All authorities on safety claim that accidents do not just happen.
   According to this statement, it follows that

    A. accidents result from the deliberate intention to do damage
    B. an accident is the result of a coincidence
    C. every accident has a definite cause
    D. it would be possible to eliminate accidents entirely

    5.____

6. A supervisor who has instructed his men in safety measures has a driver who got hurt as a result of violating a safety rule six months after the instruction period.
   Of the following, the CORRECT procedure to follow is to

    A. assume that since the instructions were given originally, that should be sufficient
    B. caution drivers working under dangerous conditions continually
    C. reprimand the driver since the man should be responsible for his own safety
    D. understand the driver's attitude since other drivers probably also ignore safety rules

    6.____

7. Records have been kept by insurance companies showing that when a supervisor who has a good safety record goes into another plant of his own company which previously had not had as good a safety record, he is able by his supervision to bring down the number of accidents in that plant. It has also been found that supervisors who were not good on safety when transferred to another plant which had a good record were not able to maintain good safety standards.
According to this paragraph,

   A. a supervisor with a good safety record if put in a plant with a bad safety record has more work to do
   B. a foreman with a bad safety record on moving to a plant with a good safety record can take it easy
   C. safety-minded supervisors make their influence felt among their subordinates
   D. the safety record in one plant is not related to the record in another plant

8. Of the following instructions relative to safety, the one which is NOT the specific duty of the supervisor to enforce is:

   A. Always sweep against traffic
   B. Do not back up without proper guidance
   C. Do not bear out from the curb in a truck without signaling
   D. Get a truck to replace a damaged one

9. In administering first aid to the victim of an accident who is suffering from more than one of the following conditions, you should treat FIRST for

   A. asphyxiation B. fractured bones
   C. poisoning D. profuse arterial bleeding

10. In treating a deep cut in the leg which results in severe bleeding, the FIRST procedure to follow is to

    A. apply a tourniquet between the wound and the heart
    B. apply digital pressure directly on the cut
    C. give the patient a strong stimulant
    D. keep the victim in a prone position, head elevated

11. A laborer has cut a small gash in his leg on the edge of a jagged timber.
Of the following, the BEST one to use directly on the wound is

    A. adhesive tape B. collodion
    C. friction tape D. a sterile gauze pad

12. Of the following, the one which is NOT used to control bleeding is

    A. cold applications B. digital pressure
    C. hot applications D. tourniquets

13. Of the following statements concerning safety precaution in a garage, the LEAST accurate is:

    A. All machinery for cutting, drilling, and charging should have the proper safeguards attached
    B. Better lighting will often eliminate dangerous working positions

C. Hazardous conditions should be corrected as soon as they are discovered
D. Principles of accident prevention differ in small, as compared with large, garages

14. Concerning accidents, the MOST accurate statement is:

    A. All individuals are equally prone to have accidents
    B. It is probable that most of the accidents occur to the same few persons in any one group
    C. Pessimistic people tend to be more careful and so have fewer accidents
    D. Physiological and psychological factors have little effect on the control of accidents

15. Of the following methods of cleaning grease and oil from slippery floors, the LEAST desirable is to

    A. clean the grease off the floor by washing with gasoline
    B. cover the floor with air slack lime, allow it to remain for several hours, and then scrape the floor
    C. scrape the floor and remove grease with caustic soda or potash
    D. sprinkle common sand and leave on the floor for a while, then use a stiff broom to sweep away

16. Of the following statements concerning a carbon dioxide extinguisher, the LEAST accurate is:

    A. An operator of such equipment needs no safeguards in an unventilated room
    B. These extinguishers are effective on electrical equipment fires
    C. Such extinguishers do not have to be protected against freezing
    D. These extinguishers should be held so that their discharge is aimed at the base of the flame

17. A supervisor who is asked by a citizen about the Department's procedures should

    A. give the citizen as little information as possible
    B. refer the citizen to the Department's public relations people or to the Commissioner
    C. refuse to discuss the matter since it might concern higher echelons of the Department
    D. tell the citizen as much factual information as possible

18. Of the following, the one which is the LEAST important reason for collecting rubbish is that it

    A. assists in the elimination of fire hazards
    B. helps in the extermination of vermin
    C. prevents health hazards of decayed vegetable matter
    D. rids the streets of trash piles

19. In the destructor method of garbage disposal, the LEAST necessary of the following requirements as to the location of the incinerator is that

    A. it should be centrally located with respect to the population it serves
    B. the site should be large enough so that noise and odors are isolated
    C. the location of the plant should be such as to dissipate possible odors
    D. there should be adequate area for garbage trucks to unload

20. Of the factors listed, the LEAST important in setting up a collection route is the 20._____
    A. average output of waste
    B. daily weather variations
    C. disposal facilities and their distance from the section
    D. necessity for daily or tri-weekly service

KEY (CORRECT ANSWERS)

1.	D	11.	D
2.	B	12.	C
3.	A	13.	D
4.	A	14.	B
5.	C	15.	A
6.	B	16.	A
7.	C	17.	D
8.	D	18.	C
9.	D	19.	A
10.	B	20.	B

TEST 3

DIRECTIONS: Each question or incomplete statement is followed by several suggested answers or completions. Select the one that BEST answers the question or completes the statement. *PRINT THE LETTER OF THE CORRECT ANSWER IN THE SPACE AT THE RIGHT.*

1. A person starting a safety campaign should be aware that

   A. accidents are more likely to occur to the same few people in an organization
   B. accidents occur equally to all kinds of people
   C. people who worry are usually more careful and, therefore, have a lower accident rate
   D. the physical and emotional condition of a person has no effect on his accident potential

   1.____

2. The one which does NOT apply to accident prevention is that the employees should

   A. always use the protective clothing and devices supplied
   B. always make reports after a serious accident has occurred
   C. avoid teasing and horseplay when on the job
   D. not use equipment when safety devices are not working

   2.____

3. During a safety campaign, all employees were taught safety rules. Three months later, a driver is hurt as a result of violating one of these rules.
The BEST procedure for the supervisor to follow is to

   A. do nothing since the employees were so recently instructed in safety
   B. ignore the situation since most drivers tend to forget and ignore safety rules
   C. reprimand the driver since a grown man should be responsible for his own safety
   D. warn drivers working under dangerous conditions frequently

   3.____

4. Accidents do not just happen.
The MOST direct conclusion from this statement is that accidents

   A. are the result of determinable causes
   B. indicate a deliberate intention to cause damage
   C. result from a chance of coincidence of events
   D. will some day be able to be entirely eliminated

   4.____

5. The one which refers LEAST to safety is:

   A. Always face traffic when sweeping
   B. Have another person direct you when backing a truck
   C. Replace a disabled truck as soon as possible to keep the work moving
   D. Signal before starting a truck away from the curb

   5.____

6. Insurance companies have gathered data showing that when a supervisor with a good safety record goes into another plant which previously had not had as good a safety record, he is able by his supervision to cut down the number of accidents in that plant. The opposite has also been proven.
According to this paragraph,

   6.____

47

A. a good record in one plant acts as and sets a high standard for the supervisor when he takes over a new plant
B. once a good record has been established, it does not take too much effort to keep it up
C. a safety-minded supervisor makes his influence felt wherever he happens to work
D. the safety record of different plants is determined more by the workers than the supervisor

7. In the event of a poison gas attack, civil defense authorities advise civilians to _____ doors and windows and go to _____.

   A. open; upper floors
   B. close; upper floors
   C. open; the basement
   D. close; the basement

8. In addition to cases of submersion, artificial respiration is a recommended first aid procedure for

   A. sunstroke
   B. chemical poisoning
   C. electrical shock
   D. apoplexy

9. An injury to a muscle or tendon brought about by severe exertion and resulting in pain and stiffness is called a

   A. strain   B. sprain   C. bruise   D. fracture

10. Of the following kinds of wounds, the one in which there is the LEAST danger of infection is a(n) _____ wound.

    A. abrasive
    B. punctured
    C. lacerated
    D. incised

11. When a person is found injured on the street, it is GENERALLY advisable, pending arrival of a physician, to help prevent fainting or shock by keeping the patient

    A. in a sitting position
    B. lying down with the head level
    C. lying down with the head raised
    D. standing on his feet

12. When an injured person appears to be suffering from shock, of the following, it is MOST essential to

    A. loosen his clothing
    B. keep him warm
    C. administer a stimulant
    D. place him in a prone position

13. Two of the CHIEF sources of traffic accidents to pedestrians in recent years were for crossing a street

    A. against the light and crossing past a parked car
    B. at a point other than the crossing and crossing against the light
    C. at a point other than the crossing and running off the sidewalk
    D. against the light and failing to observe whether cars were making right or left turns

14. You hear the cries of a boy who has fallen through the ice. 14.____
 The FIRST thing you should do is to

    A. rush to the nearest call telephone and summon paramedics
    B. call upon passersby to summon additional patrolmen
    C. rush to the spot from which the cries came and try to save the boy
    D. rush to the spot from which the cries came and question the boy concerning his identity so that you can summon his parents

15. You notice that a man is limping hurriedly, leaving a trail of blood behind him. You question him and his explanation is that he was hurt accidentally while he was watching a man clean a gun. 15.____
 You should

    A. treat his wound
    B. have him sent to the nearest hospital
    C. ask him whether the man had a license for his gun
    D. ask him to lead you to the man who cleaned his gun so that you may question him about the accident

16. You see a teenage boy dashing out of a dry cleaning store with his clothes afire. 16.____
 The BEST action for you to take in this situation is to

    A. stop the boy and roll him in a coat to smother the flames
    B. lead the boy quickly to the nearest store and douse him with large quantities of water
    C. remove all burning articles of clothing from the boy as quickly as possible
    D. take the boy back into the dry cleaning store, where a fire extinguisher will almost certainly be available to extinguish the flames quickly

17. The *Schaefer Method* of first aid is MOST helpful for 17.____

    A. stopping bleeding
    B. transporting injured persons
    C. promoting respiration
    D. stopping the spread of infection

18. You see a middle-aged man crossing the street crying out with pain and pressing his hand to his chest while standing perfectly still. You suspect that he may have suffered a heart attack. 18.____
 You should FIRST

    A. ask him to cross the street quickly in order to prevent his being hit by moving traffic
    B. permit him to lie down flat in the street while you divert the traffic
    C. ask him for the name of his doctor so that you can summon him
    D. request a cab to take him to the nearest hospital for immediate treatment

19. Which one of the following persons who has been injured in an accident should be the FIRST to be given first aid? 19.____
 A

    A. woman with a surface cut on her leg
    B. man with a sprained wrist

49

C. man with a deep cut on the cheek which is bleeding profusely
D. woman who complains that her left hand feels broken

20. When giving first aid for open wounds, the FIRST objective is to 20.____
    A. remove embedded foreign objects from inside the wound
    B. stop the bleeding
    C. clean the wound using antiseptics
    D. call the paramedics at 911

KEY (CORRECT ANSWERS)

1.	A		11.	B
2.	B		12.	B
3.	D		13.	B
4.	A		14.	C
5.	C		15.	B
6.	C		16.	A
7.	B		17.	C
8.	C		18.	B
9.	A		19.	C
10.	D		20.	B

EXAMINATION SECTION
TEST 1

DIRECTIONS: Each question or incomplete statement is followed by several suggested answers or completions. Select the one that BEST answers the question or completes the statement. *PRINT THE LETTER OF THE CORRECT ANSWER IN THE SPACE AT THE RIGHT.*

1. The BEST indicator of reduced health hazard of a product is

    A. better sales
    B. fewer product defects
    C. fewer liability claims
    D. fewer returns
    E. government approval

    1.____

2. Workers exposed to toxic agents NOT known to cause cancer are *more likely* to have

    A. recurring complex examinations
    B. recurring simple examinations
    C. single, simple examinations
    D. single, complex examinations
    E. examinations each time exposures exceed TLV values

    2.____

3. An employer must FIRST make a written exposure determination when

    A. any employee may be exposed at the action level
    B. a regulated substance is released into the workplace air
    C. employee measurement indicates exposure at the action level
    D. employee measurement indicates exposure above the action level
    E. any substance is released into the air of the workplace

    3.____

4. Responsibilities of the industrial hygienic organization are to
    I. establish hygienic standards
    II. specify the design and quality of all types of personal protective equipment
    III. review work practices
    IV. make certain employees have been examined and approved for the job
    V. plan all operations to prevent unnecessary exposure

    The CORRECT answer is:

    A. I, IV, V
    B. II, V
    C. I, III, V
    D. I, II, III
    E. II, IV

    4.____

5. Program measurement of educational activities is *least likely* to involve

    A. consumers
    B. management
    C. workers
    D. supervisors
    E. members of the community

    5.____

6. The LEAST important consideration in measuring the effectiveness of programs to control occupational health hazards is

    A. labor turnover
    B. workers' compensation costs
    C. employee morale
    D. increased productivity
    E. sickness and employee absenteeism

    6.____

7. Appendix B in the Federal Register includes
   I. physical and chemical data
   II. fire and explosion hazard data
   III. signs and symptoms
   IV. toxicology
   V. information on how to take air samples

   The CORRECT answer is:

   A. I, IV, V
   B. I, II, IV
   C. I, III, IV
   D. I, II, IV, V
   E. I, II, V

8. _____ organizations are MOST involved in teaching and training programs.

   A. Supervisor
   B. Medical
   C. Safety
   D. Industrial hygiene
   E. Engineering

9. The PRIMARY benefit of an industrial hygiene program is

   A. good employee health
   B. control of occupational disease
   C. increased productivity
   D. increased employee efficiency
   E. accident prevention

10. Adequate summary records of concentrations of toxic substances like vinyl chloride monomer shall be maintained and held available for inspection for _____ year(s).

    A. 10   B. 5   C. 20   D. 30   E. 1

11. Which of the following is NOT a minimum requirement of the Standard Completion Program's employee training program?

    A. Review signs and symptoms of exposure to regulated materials
    B. Instructions to report any sign, symptom, or medical condition to the employer
    C. Training in the safe use of regulated substances
    D. Training of emergency procedure and protective equipment
    E. Training in the details of clean-up and disposal if spills occur

12. The BEST measure of conditions affecting comfort is(are)

    A. employee complaints
    B. labor turnover
    C. employee morale
    D. fewer product defects
    E. increased productivity

13. For maximum benefit, _____ is essential to any occupational health program.

    A. maintenance of accurate attendance records for personnel
    B. pre-employment psychological profiles of employees
    C. supervision of the health status of workers by qualified medical personnel
    D. screening programs which eliminate those workers who demonstrate susceptibility
    E. mandatory physical exams for all employees on a six-month basis

14. The employer must be informed of a potentially harmful work environment detected through examination of persons subjected to it because

    A. compensation policies require full disclosure
    B. administrative controls precede all others
    C. OSHA requires employers to keep accurate records
    D. employees are employer's biggest investment
    E. employer approval is necessary before the results can be supplied to the employee's personal physician

14.____

15. The objective of the Standards Completion Program is

    A. implementation of workroom level standards
    B. the expansion and completion of existing workroom level standards promulgated by the Department of Labor
    C. to help the employer to understand why and how to be in compliance with workroom level standards
    D. to develop standards to measure the performance of the industrial hygiene program
    E. the expansion and completion of uniform monitoring techniques of workroom level standards

15.____

16. As a result of manufactured products coming under a greater number of federal requirements, the industrial hygienist has increased liaison specifically with

    I. NRC
    II. BRH
    III. OSHA
    IV. CPSC
    V. TOSCA

The CORRECT answer is:

    A. I, II, V
    B. I, III, V
    C. I, II, III
    D. II, IV, V
    E. I, II, IV

16.____

17. It is the employee's responsibility to

    A. develop and practice good habits of personal hygiene and housekeeping
    B. consult with safety professionals for aid in fulfilling his responsibilities
    C. conduct safety surveys at his work area
    D. maintain a work environment that assures maximum safety
    E. submit to a yearly physical exam

17.____

18. The _____ must consent to have the results of an occupational health program's physical exam supplied to the worker's personal physician or any other person or group.

    A. occupational physician
    B. employee
    C. industrial hygienist
    D. employer
    E. all of the above

18.____

19. When a personal monitoring discloses exposure to an employee beyond the OSHA standard, the individual should be notified in writing within _____ working days.

    A. 1-3   B. 20   C. 10   D. 5   E. 30

19.____

20. All of the following are reliable indications of improvement in environmental conditions affecting job efficiency EXCEPT

   A. increased productivity
   B. reduction in employee fatigue
   C. fewer product defects
   D. lower accident frequency rate
   E. lower labor turnover

21. Proper monitoring is NOT essential to the program measurement of

   A. environmental conditions affecting comfort
   B. health hazards of the firm's products
   C. air pollution control
   D. control of occupational health hazards
   E. environmental conditions affecting comfort

22. It is essential that outside help used for medical surveillance

   A. be familiar with conditions similar to those under which he is working
   B. have experience handling workers' compensation claims
   C. be familiar with OSHA requirements
   D. have access to available employee medical history information
   E. be familiar with the Standards Completion Program

23. Appendix C of the Federal Register is written for the

   A. employee   B. safety professional
   C. examining physician   D. employer
   E. purchasing agent

24. All three types of medical surveillance activities proposed by the Standards Completion Program require

   A. examinations be made available to the employee
   B. preplacement medical exams
   C. worker completion of a health questionnaire
   D. medical surveillance procedures be made available if and when workers develop symptoms or have acute exposures
   E. all of the above

25. The engineering organization should notify the _____ organization whenever the introduction of new operations or processes is planned.
   I. supervisory
   II. medical
   III. industrial hygiene
   IV. safety
   V. purchasing

   The CORRECT answer is:

   A. I, III, IV   B. I, IV, V   C. III, IV
   D. I, V   E. II, III, IV

KEY (CORRECT ANSWERS)

1.	C	11.	E
2.	B	12.	A
3.	B	13.	C
4.	D	14.	C
5.	A	15.	B
6.	D	16.	D
7.	E	17.	A
8.	C	18.	B
9.	A	19.	C
10.	D	20.	E

21. B
22. C
23. C
24. D
25. E

EXAMINATION SECTION
TEST 1

DIRECTIONS: Each question or incomplete statement is followed by several suggested answers or completions. Select the one that BEST answers the question or completes the statement. *PRINT THE LETTER OF THE CORRECT ANSWER IN THE SPACE AT THE RIGHT.*

1. Employer compliance with OSH Act involves

    A. control of health exposures
    B. analysis of occupational safety and health statistics
    C. enforcement of employee obligations
    D. promulgating safety and health standards
    E. all of the above

2. What duties are triggered when the action level is reached?
    I. Exposure measurement
    II. Engineering controls
    III. Medical surveillance
    IV. Employee training
    V. Work practice controls

    The CORRECT answer is:

    A. I, II, III   B. I, III, IV
    C. I, IV, V   D. II, III, IV
    E. II, III, V

3. A serious penalty may be adjusted downward by as much as ____ percent.

    A. 40   B. 50   C. 25   D. 5   E. 10

4. The Federal Mine Safety and Health Amendment Act of 1977 transfers authority, for enforcement of mining safety and health, to the

    A. Department of the Interior
    B. Department of Labor
    C. Department of Health, Education and Welfare
    D. Bureau of Mines
    E. National Institute of Health

5. Safety and health regulations and standards which have the force and effect of law are issued by the

    A. Bureau of Labor Standards
    B. Occupational Safety and Health Administration
    C. Secretary of Health, Education and Welfare
    D. Secretary of Labor
    E. Assistant Secretary for Occupational Safety and Health

6. Court authority is necessary to enforce OSH Act to
   I. inspect when no advance notice has been given
   II. inspect records of industrial injuries and illnesses
   III. invoke criminal penalties
   IV. shut down an operation
   V. appeal OSHA actions

   The CORRECT answer is:

   A. I, II, III
   B. I, III, IV
   C. II, III, IV
   D. III, IV
   E. none of the above

7. A "serious violation" is defined as a condition

   A. where there is reasonable certainty that a hazard exists that can be expected to cause death or serious physical harm
   B. that has no direct or immediate relationship to job safety or health
   C. where there is a substantial probability that death or serious physical harm could result
   D. which is responsible for a fatality or multiple hospitalization incidents
   E. where a hazard is the result of unsafe work practices

8. The ____ is authorized by The Toxic Substance Control Act to require and obtain industry-developed data on the production, use and health effects of chemical substances and mixtures.

   A. Public Health Service Administration
   B. Occupational Safety and Health Administration
   C. National Center for Toxicological Research
   D. Environmental Protection Agency
   E. National Institute of Occupational Safety and Health

9. The Toxic Substances Control Act may LOWER the need for regulation by

   A. identifying hazards at the premarketing stage
   B. making OSHA more effective
   C. guarding against interagency duplication
   D. expanding research activities
   E. not issuing detailed workplace standards

10. How many other violations (exclusive of serious violations) that have a direct relationship to job safety and health and probably would not cause death or serious physical harm, must be found before any penalty can be imposed?

    A. 3   B. 1   C. 5   D. 10   E. 20

11. The ____ is(are) responsible for coordinating the technical aspects of the health program within the region.

    A. regional compliance officers
    B. area industrial hygienist
    C. area director
    D. national OSHA office
    E. regional office industrial hygienist

12. Horizontal Standards apply to

   A. establishing standards for engineering details
   B. all workplaces and relates to broad areas
   C. specific industries
   D. specific categories of workers
   E. establishing health and safety objectives

13. The ____ licenses and regulates the use of nuclear energy to protect public health and safety and the environment.

   A. Atomic Energy Commission
   B. National Institute for Occupational Safety and Health
   C. Radiological Assistance Program
   D. Bureau of Radiological Health
   E. Nuclear Regulatory Commission

14. Reviews of decisions of contested OSHA citations are conducted by the

   A. Occupational Safety and Health Review Commission
   B. U.S. Department of Labor
   C. Office of the Area Director
   D. State Supreme Court
   E. U.S. Court of Appeals

15. The compliance officer

   A. collects health hazard information
   B. shuts down operations where conditions of "imminent danger" exist
   C. collects appropriate samples
   D. investigates health complaints
   E. conducts an industrial hygiene inspection

16. OSHA defines "action level" as

   A. parity with the permissible exposure level
   B. concentrations below maximum allowable concentrations
   C. concentrations below minimum allowable concentrations
   D. one-half the permissible exposure level
   E. a reference point for control purposes

17. Which provision of the standard are employers obligated to when no employee is exposed to airborne concentrations of a substance in excess of the action level?

   A. Special training for employees
   B. Obtaining medical history statements
   C. Measuring employee exposure
   D. All of the above
   E. None of the above

18. Which of the following has the HIGHEST priority on OSHA's schedule of inspections?

   A. Response to employee complaints
   B. Random inspections of "high hazard industries"

C. Response to community complaints
D. Response to multiple hospitalizations incidents
E. Response to reported conditions of "imminent danger"

19. A non-serious penalty has been adjusted from $1000.00 to $500.00. The LOWEST amount an employer may pay if the violation is corrected within the prescribed abatement period is

   A. $500   B. $375   C. $125   D. zero   E. $250

20. Chemicals that are exempt from premarket reporting are those
   I. produced in small quantities solely for research
   II. used for test marketing purposes
   III. determined not to present an unreasonable risk
   IV. intended for export only
   V. used exclusively for commercial purposes
   The CORRECT answer is:

   A. I, II, III
   B. I, III, IV
   C. I, III, IV, V
   D. II, III
   E. all of the above

21. The purpose of federal supervision of current state programs is to

   A. establish enforcement procedure
   B. determine excess exposures
   C. establish procedures for measuring exposure levels
   D. provide technical advice for sophisticated engineering systems
   E. achieve more uniform state inspection under federal standards

22. Which OSHA standard consists of TLVs?

   A. Performance
   B. Health
   C. Design
   D. Vertical
   E. Horizontal

23. How many federal working days of receipt of notice of the enforcement action does an employer have to contest an OSHA citation or penalty?

   A. 10   B. 15   C. 30   D. 45   E. 60

24. The PRINCIPAL federal agency engaged in research in the national effort to eliminate on-the-job hazards is the

   A. National Institute for Occupational Safety and Health
   B. Occupational Safety and Health Administration
   C. Environmental Protection Agency
   D. Food and Drug Administration
   E. Mine Safety and Health Administration

25. The area office updates the workplace inventory data

   A. semi-annually
   B. yearly
   C. every 2 years
   D. every 5 years
   E. following each inspection

KEY (CORRECT ANSWERS)

1. A
2. B
3. B
4. B
5. D

6. C
7. C
8. D
9. A
10. D

11. E
12. B
13. E
14. E
15. A

16. D
17. C
18. D
19. E
20. A

21. E
22. A
23. B
24. A
25. B

READING COMPREHENSION
UNDERSTANDING AND INTERPRETING WRITTEN MATERIAL
EXAMINATION SECTION
TEST 1

Questions 1-10.

DIRECTIONS: Each question or incomplete statement is followed by several suggested answers or completions. Select the one that BEST answers the question or completes the statement. *PRINT THE LETTER OF THE CORRECT ANSWER IN THE SPACE AT THE RIGHT.*

1. Accident prevention is an activity which depends for success upon factual information, research, and analysis. Experience has proved that all accidents can be prevented through the correct application of basic accident prevention methods and techniques determined from factual cause data. Therefore, to achieve the maximum results from any safety and health program, a uniform system for the reporting of accidents and causes is established. The procedures required for a report, when properly carried out, will determine accurate cause factors and the most practical methods for applying preventive or remedial action. According to the above paragraphs, which of the following statements is MOST NEARLY correct? 1.____

   A. No matter how much effort is put forth, there are some accidents that cannot be prevented.
   B. Accident prevention is a research activity.
   C. Accident reporting systems are not related to accident prevention.
   D. The success of an accident prevention program depends on the correct use of a uniform accident reporting system.

Questions 2-7.

DIRECTIONS: Questions 2 through 7 are to be answered ONLY according to the information given in the following accident report.

DATE: February 2

TO: Edward Moss, Superintendent
Pacific Houses
2487 Shell Road
Auburnsville, Illinois

SUBJECT: Report of Accident to
Philip Fay, Employee
1825 North 8th St.
Auburnsville, Ill.
Identification #374-24

Philip Fay, an employee, came to my office at 10:15 A.M. yesterday and told me that he hurt his left elbow. When I asked him what happened, he told me that 15 minutes ago, while shoveling the snow from in front of Building #14 at 2280 Stone Ave., he slipped on some snow-covered ice and fell on his elbow. Joseph Sanchez and Arthur Campbell, who were working with him, saw what happened.

Mr. Fay complained of pain and could not bend his left arm. I called for an ambulance right away. A police patrol car from the 85th Precinct arrived 15 minutes later, and Patrolman Johnson, Shield #8743, said that an ambulance was on the way. At 10:45 A.M., an ambulance arrived from Auburn Hospital. Dr. Breen examined Mr. Fay and told me that he would have to go to the hospital for some x-ray pictures to determine how bad the injury was. The ambulance left with Mr. Fay at 11:00 A.M.

At 3:45 P.M., Mr. Fay called from the hospital and told me that his arm had been put in a cast in the emergency room of the hospital. He was told that he had fractured his left elbow and would have to stay out of work for about four weeks. He is to report back at the hospital in three weeks for another examination and to see if the cast can be taken off. His wife was at the hospital with him, and they were now going home.

Attached are the statements from the witnesses and our completed REPORT OF INJURY form.

William Fields
Foreman

2. Which one of the following did NOT see the accident?

   A. Campbell   B. Fay   C. Fields   D. Sanchez

3. The CORRECT date and time of accident is February

   A. 2, 10:00 A.M.   B. 2, 10:15 A.M.
   C. 1, 10:00 A.M.   D. 1, 10:15 A.M.

4. The ambulance came about _____ hour after _____.

   A. 1/4 ; the accident   B. 1/4 ; it was called
   C. 1/2; the accident   D. 1/2; it was called

5. It is not possible to tell whether Fay went to report the accident right away because the report does NOT say

   A. how long it takes to get from Building #14 to the foreman's office
   B. how long it takes to get from Stone Ave. to Shell Rd.
   C. whether Fay telephoned the foreman first
   D. whether the foreman was in his office as soon as Fay got there

6. From the facts in the report, Fay's action might be criticized because he

   A. did not give the foreman the complete story of what had happened
   B. did not take Campbell or Sanchez with him when he went to the foreman's office in case he should need help on the way
   C. did not remain at the accident site and send Sanchez and Campbell to bring the foreman
   D. telephoned from the hospital and by using his arm to do this he might have aggravated his condition

7. Assuming that the report gives the complete story of this incident, the action of the foreman may be criticized because he did NOT

   A. call an ambulance soon enough
   B. go to the hospital with the ambulance and stay with the injured man until he was discharged
   C. have the injured man sign a release of claim against the department
   D. make an on-the-spot investigation of the accident scene nor take corrective action

Questions 8-10.

DIRECTIONS: Questions 8 through 10 are to be answered ONLY according to the information given in the following passage.

A foreman has four maintainers and two helpers assigned to him. Listed below are the maintainers and helpers and their rate of speed in completing the assignments given to them. Assume all the foreman's men (maintainers and helpers) are of equal technical ability but some work faster than others while some are slower in completing their assignments. In all cases, no overtime is to be granted.

Maintainer E - works at average rate of speed
Maintainer F - works at twice the rate of speed as Maintainer E
Maintainer G - works at the same rate of speed as Maintainer E
Maintainer H - works at half the rate of speed as Maintainer E
Helper J - works at same rate of speed as Maintainer G
Helper K - works at same rate of speed as Maintainer H

8. A certain job must be done immediately, and Maintainer H and Helper J are the only men available.
   If Maintainer F, working alone, could normally complete this job in six days, the TOTAL time this foreman should allot to Maintainer H and Helper J to complete the same job is _____ days.

   A. 3   B. 4   C. 8   D. 12

9. While Maintainer E and Helper J are working on a job, Helper J reports that he will be out sick for at least a week. The job normally would have taken four more days to complete, and it must be completed within these four days.
   If Maintainer H and Helper K are the only two men available, this foreman should

   A. assign Helper K to replace Helper J
   B. assign Maintainer H to replace Helper J
   C. assign both Maintainer H and Helper K to replace Helper J
   D. inform his assistant supervisor that the job cannot be completed on time

10. This foreman has assigned all six of his men to a routine maintenance job. At the end of two days, the job is four-fifths completed; and instead of reassigning all his men the following day when they would finish early, the foreman cuts the gang so that the job will take one more full day to finish. The work gang on the last day should consist of Maintainer(s)

    A. F and H
    B. F and Helper J
    C. E and Helpers J and K
    D. G and H and Helper K

Questions 11-25.

DIRECTIONS: Each question consists of a statement. You are to indicate whether the statement is TRUE (T) or FALSE (F). *PRINT THE LETTER OF THE CORRECT ANSWER IN THE SPACE AT THE RIGHT.*

Questions 11-15.

DIRECTIONS: Questions 11 through 15 are to be answered ONLY according to the information given in the following paragraph.

USING LADDERS

All ladders must be checked each day for any defects before they are used. They should not be used if there are split rails or loose rungs or if they have become shaky. Two men should handle a stepladder which is over eight feet in height, one man if the ladder is smaller. One man must face the ladder and hold it with a firm grasp while the other is working on it. When you climb a ladder, always face it, grasp the siderails, and climb up one rung at a time. You should come down the same way.

11. A ladder which is new does not have to be inspected before it is used. 11.____

12. A ladder with a loose rung may be used if this rung is not stepped on. 12.____

13. A stepladder 6 feet long may be handled by one man. 13.____

14. If a 10-foot stepladder is used, one man must hold the ladder while the other works on it. 14.____

15. The siderails of a ladder do not have to be held when climbing down. 15.____

Questions 16-20.

DIRECTIONS: Questions 16 through 20 are to be answered ONLY according to the information given in the following paragraph.

TRAFFIC ACCIDENTS

Three auto accidents happened at the corner of Fifth Street and Seventh Avenue. The first, at 7:00 P.M. last night, knocked down a light pole when two cars collided. At 8:15 A.M. this morning, two other autos crashed head on. This afternoon, at 12:30 P.M., another pair of cars crashed. One of them jumped the curb, knocked over two traffic signs, and damaged three parked cars at the corner service station. No serious injury to the drivers was reported, but all the cars involved were severely damaged.

16. Nine cars were damaged in the three accidents. 16.____

17. The three accidents happened within a period of 14 hours. 17.____

18. A service station is located at the corner of Fifth Street and Seventh Avenue. 18.____

19. In the last accident, both cars jumped the curb and knocked over two light poles. 19.____

20. The drivers of the cars in the last accident were badly hurt. 20.____

Questions 21-25.

DIRECTIONS: Questions 21 through 25 are to be answered ONLY according to the information given in the following paragraph.

LIFTING

Improper lifting of heavy objects is a frequent cause of strains and ruptures. When a heavy object is to be lifted, an employee should stand close to the object and face it squarely. The feet are spread slightly apart, and one foot is a little ahead of the other. Then, bend the knees to bring the body down to the object and keep your back comfortably vertical. Raise the object slightly to see if you can lift it alone. If you can, get a firm grasp with both hands, balance the object, and raise it by straightening the legs, but still keeping the back erect. The raising motion is gradual, not swift. In this way you use the leg muscles which are the strongest muscles in the body. This method of lifting prevents strain to the back muscles which are weak and not built for lifting purposes.

21. Many ruptures are the result of not lifting heavy objects in the correct manner.   21.____

22. When an employee lifts a heavy package, he should keep his feet close together in order to balance the load.   22.____

23. When lifting a heavy object, the back should not be bent but kept upright.   23.____

24. It is best to lift heavy objects quickly in order to prevent strains and ruptures.   24.____

25. For purposes of lifting, the leg muscles are stronger than the arm muscles.   25.____

KEY (CORRECT ANSWERS)

1.	D	11.	F
2.	C	12.	F
3.	C	13.	T
4.	D	14.	T
5.	A	15.	F
6.	B	16.	T
7.	D	17.	F
8.	C	18.	T
9.	C	19.	F
10.	B	20.	F

21. T
22. F
23. T
24. F
25. T

TEST 2

DIRECTIONS: Each question or incomplete statement is followed by several suggested answers or completions. Select the one that BEST answers the question or completes the statement. *PRINT THE LETTER OF THE CORRECT ANSWER IN THE SPACE AT THE RIGHT.*

Questions 1-8.

DIRECTIONS: Questions 1 through 8, inclusive, are based on the ladder safety rules given below. Read these rules fully before answering these questions.

LADDER SAFETY RULES

When a ladder is placed on a slightly uneven supporting surface, use a flat piece of board or small wedge to even up the ladder feet. To secure the proper angle for resting a ladder, it should be placed so that the distance from the base of the ladder to the supporting wall is one-quarter the length of the ladder. To avoid overloading a ladder, only one person should work on a ladder at a time. Do not place a ladder in front of a door. When the top rung of a ladder rests against a pole, the ladder should be lashed securely. Clear loose stones or debris from the ground around the base of a ladder before climbing. While on a ladder, do not attempt to lean so that any part of the body, except arms or hands, extends more than 12 inches beyond the side rail. Always face the ladder when ascending or descending. When carrying ladders through buildings, watch for ceiling globes and lighting fixtures. Avoid the use of rolling ladders as scaffold supports.

1. A small wedge is used to

    A. even up the feet of a ladder resting on an uneven surface
    B. lock the wheels of a roller ladder
    C. secure the proper resting angle for a ladder
    D. secure a ladder against a pole

1.____

2. An 8-foot ladder resting against a wall should be so inclined that the distance between the base of the ladder and the wall is _____ feet.

    A. 2   B. 5   C. 7   D. 9

2.____

3. A ladder should be lashed securely when

    A. it is placed in front of a door
    B. loose stones are on the ground near the base of the ladder
    C. the top rung rests against a pole
    D. two people are working from the same ladder

3.____

4. Rolling ladders

    A. should be used for scaffold supports
    B. should not be used for scaffold supports
    C. are useful on uneven ground
    D. should be used against a pole

4.____

5. When carrying a ladder through a building, it is necessary to

   A. have two men to carry it
   B. carry the ladder vertically
   C. watch for ceiling globes
   D. face the ladder while carrying it

6. It is POOR practice to

   A. lash a ladder securely at any time
   B. clear debris from the base of a ladder before climbing
   C. even up the feet of a ladder resting on slightly uneven ground
   D. place a ladder in front of a door

7. A person on a ladder should NOT extend his head beyond the side rail by more than _____ inches.

   A. 12   B. 9   C. 7   D. 5

8. The MOST important reason for permitting only one person to work on a ladder at a time is that

   A. both could not face the ladder at one time
   B. the ladder will be overloaded
   C. time would be lost going up and down the ladder
   D. they would obstruct each other

Questions 9-13.

DIRECTIONS: Questions 9 through 13 concern an excerpt of written material which you are to read and study carefully. The excerpt is immediately followed by five statements which refer to it alone. You are required to judge whether each statement

   A. is entirely true
   B. is entirely false
   C. is partly true and partly false
   D. may or may not be true but cannot be answered on the basis of the facts as given in the excerpt

It is true that in 1987 there were more strikes than in any year, excepting 1986, since 1970. However, the number of workers involved was less in 1987 than in any year since 1981, and man-days of idleness due to strikes, the MOST accurate measure of industrial strife, were less in 1987 than in any year since 1980, again excepting 1986.

9. There were fewer workers involved in strikes in 1986 than in 1981.

10. There were more strikes in 1986 than in 1987.

11. There were more strikes in 1986 than in 1970.

12. There were fewer workers involved in strikes but more man-days of idleness in 1981 than 1987.

13. There were fewer man-days of idleness and fewer workers involved in strikes in 1986 than 1987.

13._____

Questions 14-16.

DIRECTIONS: Questions 14 through 16 are to be answered on the basis of the information given in the following passage.

Telephone service in a government agency should be adequate and complete with respect to information given or action taken. It must be remembered that telephone contacts should receive special consideration since the caller cannot see the operator. People like to feel that they are receiving personal attention and that their requests or criticisms are receiving individual rather than routine consideration. All this contributes to what has come to be known as Tone of Service. The aim is to use standards which are clearly very good or superior. The factors to be considered in determining what makes good Tone of Service are speech, courtesy, understanding, and explanations. A caller's impression of Tone of Service will affect the general attitude toward the agency and city services in general.

14. The above passage states that people who telephone a government agency like to feel that they are

14._____

   A. creating a positive image of themselves
   B. being given routine consideration
   C. receiving individual attention
   D. setting standards for telephone service

15. Which of the following is NOT mentioned in the above passage as a factor in determining good Tone of Service?

15._____

   A. Courtesy   B. Education
   C. Speech   D. Understanding

16. The above passage IMPLIES that failure to properly handle telephone calls is *most likely* to result in

16._____

   A. a poor impression of city agencies by the public
   B. a deterioration of courtesy toward operators
   C. an effort by operators to improve the Tone of Service
   D. special consideration by the public of operator difficulties

Questions 17-20.

DIRECTIONS: Questions 17 through 20 are to be answered ONLY according to the information given in the following passage.

ACCIDENT PREVENTION

Many accidents and injuries can be prevented if employees learn to be more careful. The wearing of shoes with thin or badly worn soles or open toes can easily lead to foot injuries from tacks, nails, and chair and desk legs. Loose or torn clothing should not be worn near moving machinery. This is especially true of neckties which can very easily become caught in the machine. You should not place objects so that they block or partly block hallways, corridors, or other passageways. Even when they are stored in the proper place, tools, supplies, and equipment should be carefully placed or piled so as not to fall, nor have anything stick out from a pile. Before cabinets, lockers or ladders are moved, the tops should be cleared of anything which might injure someone or fall off. If necessary, use a dolly to move these or other bulky objects.

Despite all efforts to avoid accidents and injuries, however, some will happen. If an employee is injured, no matter how small the injury, he should report it to his supervisor and have the injury treated. A small cut that is not attended to can easily become infected and can cause more trouble than some injuries which at first seem more serious. It never pays to take chances.

17. According to the above passage, the one statement that is NOT true is that 17.____

    A. by being more careful, employees can reduce the number of accidents that happen
    B. women should wear shoes with open toes for comfort when working
    C. supplies should be piled so that nothing is sticking out from the pile
    D. if an employee sprains his wrist at work, he should tell his supervisor about it

18. According to the above passage, you should NOT wear loose clothing when you are 18.____

    A. in a corridor   B. storing tools
    C. opening cabinets   D. near moving machinery

19. According to the above passage, before moving a ladder you should 19.____

    A. test all the rungs
    B. get a dolly to carry the ladder at all times
    C. remove everything from the top of the ladder which might fall off
    D. remove your necktie

20. According to the above passage, an employee who gets a slight cut should 20.____

    A. have it treated to help prevent infection
    B. know that a slight cut becomes more easily infected than a big cut
    C. pay no attention to it as it can't become serious
    D. realize that it is more serious than any other type of injury

Questions 21-24.

DIRECTIONS: Questions 21 through 24 are to be answered on the basis of the following report.

TO: Thomas Smith   Date: June 14.
        Supervising Menagerie Keeper
                                                Subject:
FROM: Jay Jones
        Senior Menagerie Keeper

On June 14, a visitor to the monkey house at the zoo was noticed annoying the animals. He was frightening the animals by making loud noises and throwing stones at the animals in the cages. The visitor was asked to stop annoying the animals but did not. And he was then asked to leave the monkey house by the keeper on duty. The visitor would not leave and said that the zoo is public property and that as a citizen he has every right to be there. The keeper kept trying to pursuade the visitor to leave but was unsuccessful. The keeper finally threatened to call the police. The visitor soon left the monkey house and did not return. Fortunately, no animals were harmed in this incident.

21. The subject of the report has been left out.
Which one of these would be the BEST statement for the subject of the report?

    A. Loud noises in the monkey house
    B. Police called to monkey house
    C. Visitor annoying monkeys on June 14
    D. Monkeys unharmed by visitor

21._____

22. Which one of these is an important piece of information that should have been included in the FIRST sentence of the report?

    A. The kinds of monkeys in the monkey house
    B. Whether the visitor was a man or a woman
    C. The address of the monkey house
    D. The name of the zoo where the incident took place

22._____

23. The fourth sentence which begins with the words *And he was then asked...* is poorly written because

    A. the sentence begins with *And*
    B. the words *monkey house* should be written *Monkey House*
    C. the words *on duty* should be written *on-duty*
    D. *didn't* would be better than *did not*

23._____

24. In the sixth sentence, which begins with the words *The keeper kept trying...* , a word that is spelled wrong is

    A. trying   B. pursuade
    C. visitor   D. unsuccessful

24._____

Questions 25-27.

DIRECTIONS: Questions 25 through 27 test how well you can read and understand what you read. Read about ELEPHANTS. Then, on the basis of what you read, answer these questions.

ELEPHANTS

Elephants are peaceful animals and have very few real natural enemies. As with many other animals, when faced with danger the elephant tries to make himself look larger to his enemy. He does this by raising his head and trunk to look taller. The elephant will also extend his ears to look wider. Other threatening gestures may be made. The elephant may shift his weight from side to side, make a shrill scream, or pretend to charge with his trunk held high. If the enemy still fails to retreat, the elephant will make a serious attack.

25. When an elephant is in danger, he tries to make it appear that he is

    A. stronger   B. smaller   C. larger   D. angry

25._____

26. When he is threatened, an elephant tries to make himself look broader by 26.____

    A. taking a deep breath
    B. spreading out his ears
    C. shifting his weight from side to side
    D. holding his trunk high

27. If his enemy does not run away, the elephant will 27.____

    A. attack him
    B. run in the opposite direction
    C. hit the enemy with his trunk
    D. make a shrill scream

Questions 28-30.

DIRECTIONS: Read about PREVENTING DISEASE. Then, on the basis of what you read, answer Questions 28 through 30.

PREVENTING DISEASE

Proper feeding, housing, and handling are important in maintaining an animal's defenses against disease and parasites. The best diets are those that contain proteins, vitamins, minerals, and the other essential food elements. Proteins are especially important because they are necessary for growth. Minerals such as iron, copper, and cobalt help correct anemia. It has been shown that an animal's resistance can be decreased by improper feeding. However, it has not been proved that the use of certain types of feeds will increase the resistance of animals to infectious diseases. If animals are kept in good condition by proper diet and sanitary conditions, natural resistance to disease and parasites will be highest.

28. Food elements that are required especially for growth are 28.____

    A. minerals   B. vitamins
    C. proteins   D. carbohydrates

29. If animals are NOT fed correctly, they will 29.____

    A. have more diseases
    B. fight with each other
    C. need more proteins
    D. be able to kill parasites

30. The bodies of animals will BEST be able to fight disease naturally when they 30.____

    A. are kept warm
    B. are given immunity shots
    C. are given extra food
    D. have good diet and clean quarters

KEY (CORRECT ANSWERS)

1.	A	16.	A
2.	A	17.	B
3.	C	18.	D
4.	B	19.	C
5.	C	20.	A
6.	D	21.	C
7.	A	22.	D
8.	B	23.	A
9.	B	24.	B
10.	A	25.	C
11.	D	26.	B
12.	A	27.	A
13.	C	28.	C
14.	C	29.	A
15.	B	30.	D

EXAMINATION SECTION

TEST 1

DIRECTIONS: Each question or incomplete statement is followed by several suggested answers or completions. Select the one that BEST answers the question or completes the statement. *PRINT THE LETTER OF THE CORRECT ANSWER IN THE SPACE AT THE RIGHT.*

1. You are following up on the inspections which have been made by one of your inspectors whose work is usually satisfactory. You visit an establishment recently inspected by him and note several violations of the Code which the inspector had failed to report. You discuss the matter with the inspector who becomes highly indignant and insists that the establishment complied with the provisions of the Code at the time of his inspection.
   Under the circumstances, it would be MOST advisable for you to state that
   A. a report of the incident, submitted to the borough chief, will be included in the inspector's personnel file
   B. at the time of your visit the premises did not comply with some of the provisions of the Code
   C. future failure to report violations of the Code will be regarded as presumptive evidence of collusion
   D. you will recommend a transfer or termination of employment of the inspector if the situation occurs again

1.____

2. You have been assigned to spotcheck the daily report of an inspector. The inspector has indicated that he inspected a certain establishment at 3 P.M. The owner of the establishment insists that the inspector inspected the premises at 11 A.M.
   Of the following courses of action, you should FIRST
   A. interview the owners of the establishment visited by the inspector before and after the establishment in question
   B. secure the inspector's daily report for the previous day and check every stop on that report
   C. telephone the inspector to determine the actual time of his visit
   D. write a formal memorandum to the borough chief regarding the incident

2.____

3. You are investigating the complaint made by the owner of an establishment who alleged that an inspector spoke to him in a loud and disrespectful manner while inspecting his premises. You interview the complainant and ask him if he has any witnesses to support his complaint. He tells you that he does not. You note that the inspector found several violations of the Code in the course of his inspection of the premises.
   Under these circumstances, you should
   A. assure the owner that in the interest of good public relations the inspector involved will not be assigned to inspect the owner's premises in the future
   B. discuss the matter with the inspector before submitting your report

3.____

C. inform the owner that complaints which cannot be substantiated cannot receive further consideration
D. mark the complaint *not substantiated* and refrain from discussing it with the inspector

4. You are assigned to work with another inspector on a complex inspectional problem which has received considerable newspaper publicity. After the field work is completed, you agree to prepare a report incorporating both findings. You prepare the report and submit it directly to your superior without showing it to your fellow inspector. At a staff conference, your superior praises your report and the work performed by you; he minimizes the performance of your fellow inspector. You remain silent. Later you learn that your fellow inspector has been aggrieved by your conduct.
Of the following courses of action, it is MOST advisable that you
   A. ask the inspector whether he wishes you to try to get more credit for him
   B. discuss the matter with your fellow inspector and later with your superior to point out that the inspection was a joint effort and that your colleague should share in the credit
   C. ignore the matter and allow time to take care of the incident
   D. write a memorandum to your superior detailing precisely the work performed by the other inspector in connection with the assignment

5. Assume that you are a licensed pharmacist and would like to secure a part-time job as a pharmacist to supplement your income. In the course of your work as an inspector, you investigate an anonymous complaint against a drug store. Your investigation discloses nothing to indicate that the drug store owner has violated any provision of the Code. The owner, learning that you are a licensed pharmacist, asks you to work for him on a part-time basis.
Under the circumstances, you SHOULD
   A. *accept*, provided that the owner has not asked you for special consideration
   B. *accept*, provided that you will receive the union scale of wages
   C. *refuse* the offer since a conflict of interest situation may be involved
   D. *refuse* the offer until you have a chance to discuss it with other inspectors who are licensed pharmacists

6. Your supervisor frequently bypasses you and assigns work directly to your subordinates. You had called this matter to his attention previously. At that time, he assured you that you would not be bypassed again. However, he has continued to bypass you.
Under these circumstances, you SHOULD
   A. attempt to determine the reasons for your supervisor's action before proceeding further
   B. begin keeping a record of the instances when you are bypassed, and forward a memorandum to your supervisor's superior setting forth such instances
   C. ignore the situation until such time as your supervisor brings the matter up for discussion
   D. instruct your staff that they are to accept assignments only from you

7. In checking the daily reports of one of your inspectors, you notice that he is consistently late in beginning his working day. You discuss the matter with him and point out that disciplinary action may be taken unless he starts work promptly. He denies that he is tardy in beginning his work day. However, based on your field follow-up visits, the evidence indicates that the inspector continues to be late in starting work. Again, you discuss the matter with him and he again denies your contention that he is late in starting work.
You should
   A. again point out the need for starting work promptly and continue checking the inspector's starting time
   B. discuss the matter with other inspectors in the work group to get their advice
   C. ignore the matter as long as the inspector makes about as many inspections as others in the group
   D. report the inspector to your supervisor for appropriate disciplinary action

8. One of the inspectors in your group makes about one-fourth more inspections than any other inspector. However, his inspections do not meet satisfactory standards of quality. After you have given him training in the field, his work improves to a point where it is satisfactory. However, he still makes about one-fourth more inspections than any of the other inspectors.
You should
   A. ask that the inspector be transferred to a unit where the quantity and quality of work produced by him will be closer to the group standard
   B. ask the proprietors of establishments visited by this inspector whether the inspections were too cursory
   C. devote less time to this inspector so that you may devote more time to those inspectors who may need additional training
   D. instruct the inspector to reduce the number of his inspections to the group standard and spend more time in each establishment

9. Assume that your superior has given you an additional job to do which will require extra effort on the part of your inspectors who are now carrying a full work load. You feel that the job cannot be completed in the allotted time. You present your point of view but your superior insists that you handle the assignment without any increase in staff.
Of the following courses of action, it would be MOST advisable for you to
   A. attempt to complete the assignment within the allotted time by rescheduling and re-assigning other work
   B. commit yourself to no specific course of action while attempting to secure evidence to support your position that you should not be given the assignment
   C. insist that your superior give you some assurance that this assignment does not set a precedent for assignments of a similar nature and agree to do the job
   D. take the matter up with higher authority, preferably by memorandum, but apprise your superior of your action

10. You are conducting a conference with the inspectors assigned to you. During the conference, you make a statement regarding field inspections which you are reasonably certain is correct. One of the inspectors tells you in an offensive manner that your statement is incorrect. Some of the inspectors agree with him; others remain silent.
Under these circumstances, you SHOULD
   A. ask the inspectors who have not made any comments for their opinions and be guided by their remarks
   B. ignore the offensive manner of the speaker and state that since you are certain that you are correct, the group will be guided by your statement
   C. state that while the manner of the speaker is offensive, he is nevertheless probably correct
   D. state that you will ascertain whether your statement is correct and will advise them of it in the near future

10.____

11. Assume that you are in the habit of writing to your supervisor on subjects related to your duties. Your supervisor tells you that you are writing too many memorandums to him.
Of the following courses of action, it is MOST preferable for you to
   A. instruct your inspectors not to put in writing communications regarding the work of the unit
   B. refrain from communicating in writing with your supervisor
   C. take no notice of your supervisor's statement since the smooth functioning of an organization depends upon written communication
   D. write to your supervisor only when you feel that it is necessary

11.____

12. You are accompanying one of your recently appointed inspectors on a field inspection. His inspections take an unusually long time to complete since he is extremely meticulous.
In these circumstances, you should FIRST
   A. assure the inspector of your confidence in his ability to perform his job properly after sufficient training before criticizing his work performance
   B. seek the transfer of the inspector to a position in the department which does not require contact with the public
   C. tell the inspector that if he does not bring his work up to standard immediately, you will report him to your supervisor
   D. urge the inspector to seek employment in a field not related to his present work

12.____

13. A rumor has started among the members of our staff to the effect that you will soon be leaving government service to take a position in private industry. You know that the rumor is untrue.
You SHOULD
   A. ask your staff not to discuss matters among themselves which relate to your own affairs
   B. inform your staff that you do not intend to take a position in private industry
   C. say nothing about the matter to your staff
   D. tell your staff that you refuse to confirm or deny rumors concerning your employment prospects

13.____

14. You request an inspector to do something in a certain manner. The inspector asks you the reason for performing the operation in the manner suggested by you.
    You SHOULD
    A. change the subject of your discussion
    B. explain to the inspector that it is his job to carry out instructions—not to evaluate them
    C. give the inspector the reason for your request
    D. tell the inspector that if he thinks about the matter he will be able to determine the reason himself

    14.____

15. You are conducting a conference with your staff. One of your inspectors seems completely disinterested in the discussion.
    To get this inspector to participate, you SHOULD
    A. ask the inspector direct questions related to the subject being discussed
    B. determine if there is any subject this inspector would like the group to discuss
    C. ignore the situation until such time as the inspector shows interest
    D. tell the inspector in a polite way to pay strict attention

    15.____

16. You are conducting a conference with your staff and are having a great deal of difficulty with one of the inspectors who wants to do all of the talking. You have previously spoken privately to this inspector regarding his habit of *hogging the discussion*—to no avail.
    Under the circumstances, you SHOULD
    A. ask the inspector to act as an auditor only during conferences
    B. elicit discussion by direct questioning of other members of the staff
    C. refrain from looking at the inspector when you ask a question; this will make it impossible for him to *get the floor*
    D. tell the inspector to remain silent or to leave the group

    16.____

17. You telephone one of your inspectors, assigning him to the central office for a period of two days to perform clerical duties. The inspector complains loudly, tells you that he dislikes clerical work and that he is being treated unfairly since there are inspectors in other boroughs who are assigned less frequently to clerical duties. You explain the situation as best you can but the inspector continues to object.
    Under these circumstances, you SHOULD
    A. ask the inspector to disregard the assignment pending your inquiry into practices followed in other boroughs in assigning personnel to clerical duties
    B. promise the inspector that in the future you will do your best to give clerical assignments to people who do not voice objections to such assignments
    C. tell the inspector to report for duty in accordance with your instructions
    D. tell the inspector to take the matter up with your superior

    17.____

18. In developing an on-the-job training program for inspectors, the FIRST thing which should be determined is
    A. areas in which training is needed
    B. how many inspectors are interested in training
    C. how much will the training program cost
    D. what training aids and facilities are available

19. Assume that a recently appointed and inexperienced inspector is given a difficult assignment. He is not given any specific instructions as to how the assignment should be carried out.
    Such action is
    A. *good*; a new employee needs to be encouraged to exercise his own initiative
    B. *good*; a new employee will remember longer if he learns by himself
    C. *poor*; newly appointed employees usually need guidance
    D. *poor*; the cost of training varies from employee to employee

20. A number of important changes have taken place in several sections of the Code. You are to inform a group of inspectors of these changes and how they are to be implemented. While speaking to the group concerning the changes, one of the inspectors whom you know to be a quick learner complains that you are proceeding too slowly; another inspector whom you know to be the slowest learner in the group tells you that your teaching pace is just right.
    You SHOULD
    A. bring the session to a halt and instruct group members on an individual basis
    B. proceed at a faster rate
    C. proceed at a faster rate but allow more time for breaks
    D. proceed at the same rate

21. As part of a new inspector's training, you observe him as he conducts an inspection. The inspector completes the *score-card* on which he lists certain violations of the Code. You look at the *score-card* and note that although the inspector spoke to the establishment owner about a certain violation, the inspector failed to list the violation on the *score-card*.
    Of the following, the MOST desirable way of pointing out this omission is to
    A. ask the inspector to look at the score-card to see if anything is missing
    B. criticize the inspector in a forthright manner and impress upon him the importance of the probationary period
    C. show the score-card to the owner and ask the owner to indicate the violation which was noted but not recorded
    D. tell the inspector in the presence of the owner to list the violation and make a separate note of the omission for service rating purposes

22. Assume that you are in the field training a recently appointed inspector in inspectional techniques.
    The inspections demonstrated by you SHOULD be of the kind
    A. consistent with the high standards of experienced inspectors

B. performed by the average beginning inspector so as not to unduly discourage the trainee
C. performed by the sanitarian who barely meets the minimum acceptable standard
D. which varies sharply from one inspection to the next so that the new inspector will be able to familiarize himself with various ways in which inspections may be conducted

23. The one of the following which LEAST describes the function of planning at the senior inspector level of supervision is deciding
    A. *how* something should be done
    B. *what* must be done
    C. *who* should do it
    D. *why* something should be done

24. For effective management, delegation of responsibility MUST be accompanied by *appropriate*
    A. authority
    B. commendation
    C. compensation
    D. privilege

25. The one of the following which is LEAST a *staff* function in an organization is
    A. advising
    B. directing
    C. observing
    D. planning

26. Assume that an employee is responsible to two supervisors of equal rank for the proper performance of his duties.
    The principle of good management which is NOT being complied with is
    A. delegation of authority
    B. fixed responsibility
    C. homogeneous assignment
    D. unity of command

27. Where low morale is responsible for low work output, the FIRST step which should be taken is to
    A. determine the reason for the poor state of morale by interviewing supervisors and employees who are directly affected
    B. have the head of the organization deliver an inspirational talk to those responsible for the low work output, stressing the mission of the organization and the importance of the work involved
    C. lower standards of production to equal work output and then gradually increase these standards to the desired level
    D. withdraw privileges with regard to the granting of leave, coffee breaks, and choice of lunch hours until work output rises to a satisfactory level

28. The one of the following which is NOT usually a need which gives rise to a work simplification program in government is the need to
    A. make the job as pleasant as possible for employees
    B. make things more convenient for members of the public
    C. produce a greater quantity and higher quality of work
    D. provide additional employment in times of recession

29. The MOST practical control the inspector has over the contractor when the inspector is not satisfied with the quality of the work is to
    A. discuss withholding payment on that part of the work that is unsatisfactory
    B. threaten to have the contractor thrown off the job
    C. request that the contractor fire the men responsible for the unsatisfactory work
    D. call the owner of the company and explain the situation to him

30. In the absence of a formal training program for inspectors, the BEST of the following ways to train a new man who is to do inspection work is to
    A. give him literature on the subject so that he can learn what he has to know
    B. have him accompany an inspector as the inspector does his work so that he can learn by observing
    C. assign him the job and let him learn on his own
    D. tell him to go to a school at night that specializes in this field so that he will gain the necessary background

KEY (CORRECT ANSWERS)

1.	B	11.	D	21.	A
2.	A	12.	A	22.	A
3.	B	13.	B	23.	D
4.	B	14.	C	24.	A
5.	C	15.	A	25.	B
6.	A	16.	B	26.	D
7.	D	17.	C	27.	A
8.	C	18.	A	28.	D
9.	A	19.	C	29.	A
10.	D	20.	D	30.	B

TEST 2

DIRECTIONS: Each question or incomplete statement is followed by several suggested answers or completions. Select the one that BEST answers the question or completes the statement. *PRINT THE LETTER OF THE CORRECT ANSWER IN THE SPACE AT THE RIGHT.*

1. Assume that you are a supervisor newly assigned to a squad of inspectors. In order to establish a favorable working atmosphere, it is BEST to
   A. discipline ineffective members of your squad at regular intervals
   B. speak extensively on job-related subjects
   C. give advice on personal matters
   D. recognize and accept ideas submitted by members of your squad

1._____

2. Assume that you are a supervisor who has in his squad an ambitious inspector studying for promotion. This man takes every opportunity to as you questions about your job.
   Under the circumstances, it is BEST for you to
   A. remind him firmly that he already has a full-time job and that if he wishes to study for promotion he should do it off-duty by himself
   B. plan your time so that you can assist in his promotional aspirations
   C. tell him that you would like to help but that you do not wish to give him an advantage over others
   D. resist instructing him because if he is promoted you will lose a valuable man, thereby weakening your squad

2._____

3. A signed written complaint has been mailed directly to you alleging that one of your inspectors has been overly aggressive in that he pushed the complainant. The inspector is a good worker, and this is the first complaint ever recorded against him.
   Under the circumstances, it is BEST to
   A. notify informally the accused inspector of the nature of the complaint, and suggest that he guard his behavior in the future
   B. ignore the complaint as being too vague to warrant action
   C. have the complainant carefully investigated to see whether he has made similar complaints in the past
   D. have the complaint investigated by someone disinterested in the outcome of the matter

3._____

4. Assume that you are a supervisor in charge of an inspector who has a good work record but who, for the first time, exhibits symptoms of drunkenness. When confronted, he denies that he ever drinks and says that his apparently intoxicated behavior is really the result of his doctor's medication for the flu.
   Under the circumstances, it is BEST to
   A. ignore the situation for the present but later report the matter to your superiors
   B. tell the man that you know he's untruthful but that, because of his previous good record, you are willing to overlook his condition this time

4._____

2 (#2)

    C. accept the man's explanation, send him home for the day on sick leave, but watch for future symptoms of possible drunken behavior
    D. reprimand the man, send out for coffee to sober him up, and warn him the next time he exhibits drunken symptoms he will face severe disciplinary action

5. Assume that you are in charge of a squad of inspectors. One inspector has been performing ineffectively, although working hard. All attempts to improve his performance have failed. He is nearing the end of his probationary period. In the circumstances, it is BEST to
    A. reschedule assignments so that the rest of the squad takes over a greater share of the work load
    B. recommend separation on the ground that improvement cannot be achieved
    C. assign only the simplest cases to the man
    D. leave the man alone, since he seems to be doing the best he can

6. As a supervisor, you have been instructed by your superiors to install a radically revised system of procedure for your squad. You are concerned that your subordinates may resist the change.
The BEST way for you to secure the willing cooperation of your squad in effecting the change is to
    A. secure the participation of all your subordinates in planning for the change, emphasizing the absence of any threat to their security
    B. *sell* your subordinates on the new procedure by emphasizing that the procedure has the full backing of your superiors
    C. warn your subordinates not to sabotage the change, emphasizing that willful interference with the change will be followed by severe corrective disciplinary action
    D. appeal to your subordinates' loyalty to the agency and to yourself, emphasizing that *one hand washes the other*

7. Supervising inspectors are involved in the decision-making process.
Effective decision-making means MOST NEARLY
    A. compromising, since all decisions involve compromise
    B. selecting the course of action with the least unexpected consequences
    C. holding off on any action until circumstances dictate one particular approach
    D. securing employee participation in the planning and policy process

8. Assume that you are a supervisor in charge of an inspector who may be abusing sick leave.
Under the circumstances, the FIRST thing you should do is to
    A. interview the inspector to find out what is wrong
    B. maintain a calendar of sick leave used by the inspector to see whether a pattern develops indicating abuse
    C. warn the inspector against any further malingering
    D. institute corrective disciplinary action the very next time the inspector reports sick

9. Supervision is a social relationship. It is both the art of being a leader and a subordinate.
   This statement implies that
   A. the supervisory relationship involves an expectation of obedience on the part of the supervisor and a willingness to obey on the part of the subordinate
   B. the really successful supervisor always knows that his subordinates understand him, and doesn't have to clarify and explain his orders
   C. in the supervisory relationship, supervisor and subordinate should strive to be as friendly with each other as possible
   D. the really wise subordinate knows his job and sees to it that his supervisor knows that he knows his job

10. As a supervising inspector, you have a man in your squad who avoids difficult tasks on the ground that he cannot do the more difficult work. You have informally condoned his practice because he is effective and busy on lesser tasks, overall squad production is satisfactory, and no one has complained. Nevertheless, you decide to review the situation.
    Solely on the basis of the information presented, the LEAST effective response to this situation is to
    A. denounce the man before the group and ask for their advice on handling the matter
    B. insist on a basic work capability for all members of the squad
    C. continue the present practice informally, so long as production and morale are unaffected
    D. remind the man that professional recognition awaits those who work hard on a variety of tasks

11. Assume that you are a new supervisor in charge of a squad of inspectors. Your superior informs you that the squad has long been declining in effectiveness. Your job is to increase production without changing personnel.
    Of the following, the MOST important information for you to have in order to effect change is
    A. the reason for the squad's past production successes
    B. an accurate account of your squad's present state of mind
    C. a knowledge of the interplay of psychic needs and neighborhood surroundings in producing the squad's laxity
    D. a case history on every individual so that you can estimate the personal impact of prospective changes

12. Assume that you have become the supervisor of a high morale squad of inspectors, all of whom ae experienced and productive.
    The BEST supervisory approach for you to take to insure the continuance of an efficient squad is to
    A. leave them alone, since it doesn't pay to tinker with a well-running mechanism
    B. develop a close personal relationship with the most experienced member of your squad and use this relationship to govern the rest of the squad

C. take charge immediately, and let them know who's in charge since everything usually runs well when persons are alert
D. work problems out together, on the theory that things usually run well when the supervisor successfully seeks to build power with, rather than hold authority over, his work group

13. One of the things a supervising inspector should AVOID doing is
    A. answering unimportant questions asked by the public
    B. talking to people he does not know
    C. blaming his supervisors for all the unpleasant orders the supervising inspector must issue
    D. showing an interest in public problems

14. An angry building owner complains loudly to you, the supervisor, about the actions of the inspectors assigned to you.
    You should
    A. try to find excuses for your men's actions
    B. speak to him in the same tone of voice he is using
    C. insist that the actions of your men are correct
    D. try to answer his complaint quietly

15. In dealing with the general public, an inspector should remember that
    A. every person is an individual who may think for himself
    B. all people tend to think alike
    C. most people think alike
    D. it is best to change the public's way of thinking to what the department requires

16. An inspector is performing his job in the BEST manner when he
    A. continually checks with his supervisor to make sure each inspection is being done properly
    B. knows enough to overlook minor violations that have a negligible effect on overall
    C. varies the rules when he feels they do not meet the conditions of the job
    D. is careful and observant in his inspections

17. An IMPORTANT characteristic of a good supervisor is his ability to
    A. be a stern disciplinarian
    B. put off the settling of grievances
    C. solve problems
    D. find fault in individuals

18. At the time you hand out a job assignment, an inspector feels that he cannot complete the job within the time limit you have given him.
    You would expect the inspector FIRST to
    A. make as many inspections as possible and then report to you
    B. compare his workload to that of the other inspectors
    C. complete the work by putting in overtime before notifying you of the problem
    D. request assistance in doing the work

19. A new supervising inspector will BEST obtain the respect of the men assigned to him if he
    A. makes decisions rapidly and sticks to them regardless of whether they are right or wrong
    B. makes decisions rapidly and then changes them just as rapidly if the decisions are wrong
    C. does not make decisions unless he is absolutely sure that they are right
    D. makes his decisions after considering carefully all available information

20. A newly-appointed inspector is operating at a level of performance below that of the other employees.
    In this situation, a supervisor should FIRST
    A. lower the acceptable standard for the new inspector
    B. find out why the new inspector cannot do as well as the others
    C. advise the new inspector he will be dropped from the payroll at the end of the probationary period
    D. assign another new inspector to assist the first inspector

21. Assume that you have to instruct a new inspector on a specific departmental operation. The new man seems unsure of what you have said.
    Of the following, the BEST way for you to determine whether the man has understood you is to
    A. have the man explain the operation to you in his own words
    B. repeat your explanation to him slowly
    C. repeat your explanation to him using simpler wording
    D. emphasize the important parts of the operation to him

22. A supervising inspector realizes that he has taken an instantaneous dislike to a new inspector assigned to him.
    The BEST course of action for this supervisor to take in this case is to
    A. be especially observant of the new inspector's actions
    B. request that the new inspector be reassigned
    C. make a special effort to be fair to the new inspector
    D. ask to be transferred himself

23. A supervisor gives detailed instructions to his inspectors as to how a certain type of job is to be done.
    One ADVANTAGE of this practice is that this will
    A. result in a more flexible operation
    B. standardize operations
    C. encourage new men to learn
    D. encourage initiative in the men

24. Of the following, the one that would MOST likely be the result of poor planning is:
    A. Omissions are discovered after the work is completed.
    B. During the course of normal inspection, a meter is found to be unaccessible.
    C. An inspector completes his assignments for that day ahead of schedule.
    D. A problem arises during an inspection and prevents an inspector from completing his day's assignments.

25. Of the following, the BEST way for a supervisor to maintain good morale among his inspectors is for the supervisor to
    A. avoid correcting an inspector when he makes mistakes
    B. continually praise an inspector's work even when it is of average quality
    C. show that he is willing to assist in solving the inspector's problems
    D. accept the inspector's excuses for failure even though the excuses are not valid

KEY (CORRECT ANSWERS)

1.	D		11.	B
2.	B		12.	D
3.	D		13.	C
4.	C		14.	D
5.	B		15.	A
6.	A		16.	D
7.	B		17.	C
8.	A		18.	D
9.	A		19.	D
10.	A		20.	B

21. A
22. C
23. B
24. A
25. C

EXAMINATION SECTION
TEST 1

DIRECTIONS: Each question or incomplete statement is followed by several suggested answers or completions. Select the one that BEST answers the question or completes the statement. *PRINT THE LETTER OF THE CORRECT ANSWER IN THE SPACE AT THE RIGHT.*

1. Assume you are supervising a group of investigators. Your unit is assigned a rush job requiring a special skill and overtime work.
   Of the following, the MOST appropriate method of choosing the investigator to do this job is to

   A. assign the investigator who has the special skill required for the job
   B. ask an investigator who has previously indicated a willingness to work overtime
   C. call for a volunteer to perform this work
   D. offer the job to the investigator who is next in line to work overtime

2. Formal training programs can help remedy specific problems in an investigative unit. The one of the following that is NOT an intended result of such training programs is to

   A. eliminate the need for on-the-job training for new investigators
   B. help reduce the amount of overtime paid
   C. minimize the number of grievances made by investigators
   D. develop a pool of trained investigators needed for agency expansion

3. Periodic evaluation of subordinates' performance on the job serves all of the following purposes EXCEPT to

   A. point out weaknesses in performance to subordinates so that attempts can be made to eliminate them
   B. identify capable subordinates and insure that they are promoted
   C. indicate those subordinates who deserve training for greater responsibilities
   D. identify those subordinates who have exceptional ability

4. All of the following are proper objectives in the investigation of outside complaints about agency personnel EXCEPT the protection of the

   A. integrity and reputation of the staff
   B. public interest in identifying wrongdoers
   C. organization from liability resulting from unjust claims
   D. accused employees from disciplinary action

5. Assume that one of your subordinates had had a minor accident while performing a surveillance. In spite of your repeated demands, the subordinate refuses to prepare an accident report because he was only slightly injured. Of the following actions, it would be BEST in this situation for you to

   A. contact your superior to discuss disciplinary action
   B. have the employee file an affidavit absolving you of any responsibility for his injury
   C. ask the employee to submit a doctor's note to you on the extent of his injury
   D. call a meeting of subordinate personnel to discuss this situation

6. The one of the following that is likely to provide subordinates with the GREATEST satisfaction on the job is

   A. compensation for overtime production
   B. challenging and interesting work
   C. compensation proportional to the amount of work produced
   D. minimum responsibility for the completion of work

7. An employee is GENERALLY considered guilty of insubordination when he

   A. refuses to obey a supervisor's order with which he disagrees
   B. declines to carry out a directive he genuinely believes will cause him personal injury
   C. uses foul or abusive language among other work group members
   D. reports to work late after being warned not do do so

8. The one of the following that is GENERALLY characteristic of the more effective supervisors is that they

   A. specify every detail of the work to be done
   B. give subordinates leeway in the methods they use to complete their work
   C. supervise more closely than the less effective supervisors
   D. tend to be production-centered rather than employee-centered

9. If workers participate in planning, making important decisions, and the like, the supervisor will lose prestige and his authority will deteriorate.
   This statement is

   A. *true* because people have little respect for a leader who seeks their advice
   B. *true* because a supervisor must establish a firm command over his subordinates to be effective
   C. *false* because a skillful supervisor works with his subordinates to establish a goal and then works to reach it
   D. *false* because a supervisor gains prestige only by making all important decisions himself

10. As a supervisor, you note that while one of your subordinates does what he is told to do, he seems disinterested and lacks motivation in performing his work.
    Of the following, the BEST action for you to take to motivate this employee would be to

    A. transfer him to a more active unit
    B. give him less desirable work
    C. give him more responsibility
    D. assign him to work with a more experienced employee

11. Newly appointed supervisors will often assume responsibility for work which could be handled by their subordinates.
    Of the following, the MOST likely result of such a practice would be that the

    A. supervisor will gain the confidence of his subordinates
    B. subordinates' sense of initiative and responsibility will diminish
    C. supervisor will note an increase in the job satisfaction of his subordinates
    D. subordinates will have more time to learn more complex job skills

12. In order to accomplish the work of his unit MOST effectively, a supervisor of investigators should

    A. do the important work himself
    B. assign complete responsibility for the completion of work only to his more productive subordinates
    C. judiciously delegate authority to make decisions to his subordinates
    D. give sensitive and responsible work only to his most competent investigators

13. Assume, as a supervisor, you are approached by one of the investigators in your unit with what you consider to be a minor grievance.
    Of the following, the BEST way to handle this situation is to

    A. postpone taking any action since the passage of time usually resolves minor grievances
    B. try to resolve the problem immediately before it gets out of hand
    C. tell the investigator not to be concerned with minor grievances
    D. thank the investigator for calling the grievance to your attention and await further developments

14. Following are three guidelines a supervisor might follow in handling criticism by a superior:
    I. Consider the source of criticism before taking any action.
    II. Try to react calmly to criticism that is not justified.
    III. Analyze carefully only the criticism that requires a response.
    Which of the following CORRECTLY classifies the above guidelines into those which are valid and those which are not valid?

    A. I is valid, but II and III are not.
    B. I and II are valid, but III is not.
    C. II and III are valid, but I is not.
    D. III is valid, but I and III are not.

15. Assume that a supervisor notices that several of his subordinates, who are normally punctual, have been late for work quite often during the last few months.
    Which one of the following actions should the supervisor take FIRST in dealing with this problem?

    A. Refer the matter to the personnel staff of his agency.
    B. Schedule counseling sessions on the need for being prompt.
    C. Review his own supervision to determine whether it has been adequate.
    D. Inform the subordinates that exact records of their latenesses are being kept.

16. Following are three statements concerning principles of delegation:
    I. Supervisors should not be held accountable for work that has been delegated to their subordinates.
    II. Subordinates should normally have only one line supervisor.
    III. When subordinates are given authority that is limited by factors such as departmental rules, their responsibility is also limited.
    Which of the following BEST classifies the above statements into those that are valid and those that are not valid?

A. I is valid, but II and III are not.
B. II is valid, but I and III are not.
C. I and II are valid, but III is not.
D. II and III are valid, but I is not.

17. Following are six steps that should be taken in the course of report preparation: 17._____
    I. Outlining the material for presentation in the report.
    II. Analyzing and interpreting the facts
    III. Analyzing the problem
    IV. Reaching conclusions
    V. Writing, revising, and rewriting the final copy
    VI. Collecting data

    According to the principles of good report writing, the CORRECT order in which these steps should be taken is

    A. VI, III, II, I, IV, V
    B. III, VI, II, IV, I, V
    C. III, VI, II, I, IV, V
    D. VI, II, III, IV, I, V

18. Following are three statements concerning written reports: 18._____
    I. Clarity is generally more essential in oral reports than in written reports.
    II. Short sentences composed of simple words are generally preferred to complex sentences and difficult words.
    III. Abbreviations may be used whenever they are customary and will not distract the attention of the reader.

    Which of the following choices CORRECTLY classifies the above statements into those which are valid and those which are not valid?

    A. I and II are valid, but III is not valid.
    B. I is valid, but II and III are not valid.
    C. II and III are valid, but I is not valid.
    D. III is valid, but I and II are not valid.

19. In order to produce a report written in a style that is both understandable and effective, an investigator should apply the principles of unity, coherence, and emphasis. The one of the following which is the BEST example of the principle of coherence is   19._____

    A. interlinking sentences so that thoughts flow smoothly
    B. having each sentence express a single idea to facilitate comprehension
    C. arranging important points in prominent positions so they are not overlooked
    D. developing the main idea fully to insure complete consideration

20. Following are three statements concerning public relations in a city agency: 20._____
    I. Public relations in an agency should be the sole responsibility of a trained public relations professional
    II. Public relations involves every contact the agency has with the public, whether the contact is in person or by letter or telephone
    III. The public should be told by the agency what it is going to do and how it is going to do it before hearing a version from other sources which may be distorted

    Which of the following choices CORRECTLY classifies the above statements into those which are correct and those which are not?

A. I and II are correct, but III is not.
B. I is correct, but II and III are not.
C. II is correct, but I and III are not.
D. II and III are correct, but I is not.

21. Communication, both written and oral, is essential to the functioning of any organization. Written communication is generally more appropriate than oral communication when the information being transmitted 21._____

    A. concerns a small group of people
    B. has long-term significance
    C. is only of minimal importance
    D. is concise and simple to comprehend

22. Subordinates are MOST likely to accept changes in their work plans and schedules when their supervisor 22._____

    A. advises them that such changes must be implemented because they have been ordered by management
    B. gives them some background to help them understand the need for the changes
    C. tells them that even though he disagrees with the changes, they must be adhered to
    D. informs them he will follow up to determine how effective such changes are

Questions 23-25.

DIRECTIONS: Below is a report consisting of 15 numbered sentences, some of which are not consistent with the principles of good report writing. Questions 23 through 25 are to be answered SOLELY on the basis of the information contained in the report and your knowledge of investigative principles and practices.

To: Tom Smith, Administrative Investigator
From: John Jones, Senior Investigator

1. On January 7, I received a call from Mrs. H. Harris of 684 Sunset Street, Brooklyn.
2. Mrs. Harris informed me that she wanted to report an instance of fraud relating to public assistance payments being received by her neighbor, Mrs. I. Wallace.
3. I advised her that such a subject would best be discussed in person.
4. I then arranged a field visitation for January 10 at Mrs. Harris' apartment, 684 Sunset Street, Brooklyn.
5. On January 10, I discussed the basis for Mrs. Harris' charge against Mrs. Wallace at the former's apartment.
6. She stated that her neighbor is receiving Aid to Dependent Children payments for seven children, but that only three of her children are still living with her.
7. In addition, Mrs. Harris also claimed that her husband, whom she reported to the authorities as missing, usually sees her several times a week.
8. After further questioning, Mrs. Harris admitted to me that she had been quite friendly with Mrs. Wallace until they recently argued about trash left in their adjoining hall corridor.
9. However, she firmly stated that her allegations against Mr. Wallace were valid and that she feared repercussions for her actions.

10. At the completion of the interview, I assured Mrs. Harris of the confidentiality of her statements and that an attempt would be made to verify her allegations.
11. However, upon presentation of official identification, Mrs. Wallace refused to admit me to her apartment or grant an interview.
12. As I was leaving Mrs. Harris' apartment, I noticed a man, aged approximately 45, walking out of Mrs. Wallace's apartment.
13. I followed him until he entered a late model green Oldsmobile Cutlass, license plate #238DAB, and sped away.
14. On January 15, I returned to 684 Sunset Street, having determined that Mrs. Wallace is receiving assistance as indicated by Mrs. Harris.
15. I am, therefore, referring this matter to you for further instructions.

                                                     John Jones
                                                   Senior Investigator

23. The one of the following that indicates the MOST logical order for statements 11 through 15 is

    A. 11, 12, 13, 14, 15   B. 13, 14, 11, 12, 15
    C. 11, 13, 14, 12, 15   D. 12, 13, 14, 11, 15

24. Which of the following sentences from the report is ambiguous?
Sentence

    A. 2   B. 7   C. 8   D. 9

25. Of the following, based on the above report and your knowledge of investigative practice, it is MOST likely that investigator Jones failed to obtain the desired information from Mrs. Wallace because

    A. she was aware of Mrs. Harris' allegations
    B. she was fearful of personal injury
    C. he was not operating under cover
    D. he had not made a prior arrangement for the visit

KEY (CORRECT ANSWERS)

1.	A	11.	B
2.	A	12.	C
3.	B	13.	B
4.	D	14.	B
5.	A	15.	C
6.	B	16.	B
7.	A	17.	B
8.	B	18.	C
9.	C	19.	A
10.	C	20.	D

21. B
22. B
23. D
24. B
25. C

EXAMINATION SECTION
TEST 1

DIRECTIONS: Each question or incomplete statement is followed by several suggested answers or completions. Select the one that BEST answers the question or completes the statement. *PRINT THE LETTER OF THE CORRECT ANSWER IN THE SPACE AT THE RIGHT.*

1. Although some kinds of instructions are best put in written form, a supervisor can give many instructions verbally.
   In which one of the following situations would verbal instructions be MOST suitable?
   A. Furnishing an employee with the details to be checked in doing a certain job
   B. Instructing an employee on the changes necessary to update the office manual used in your unit
   C. Informing a new employee where different kinds of supplies and equipment that he might need are kept
   D. Presenting an assignment to an employee who will be held accountable for following a series of steps

   1.____

2. You may be asked to evaluate the organization structure of your unit.
   Which one of the following questions would you NOT expect to take up in an evaluation of this kind?
   A. Is there an employee whose personal problems are interfering with his or her work?
   B. Is there an up-to-date job description for each position in this section?
   C. Are related operations and tasks grouped together and regularly assigned together?
   D. Are responsibilities divided as far as possible, and is this division clearly understood by all employees?

   2.____

3. In order to distribute and schedule work fairly and efficiently, a supervisor may wish to make a work distribution study. A simple way of getting the information necessary for such a study is to have everyone for one week keep track of each task doe and the time spent on each.
   Which one of the following situations showing up in such study would MOST clearly call for corrective action?
   A. The newest employee takes longer to do most tasks than do experienced employees.
   B. One difficult operation takes longer to do than most other operations carried out by the section.
   C. A particular employee is very frequently assigned tasks that are not similar and have no relationship to each other.
   D. The most highly skilled employee is often assigned the most difficult jobs.

   3.____

4. The authority to carry out a job can be delegated to a subordinate, but the supervisor remains responsible for the work of the section as a whole.
As a supervisor, which of the following rules would be the BEST one for you to follow in view of the above statement?
   A. Avoid assigning important tasks to your subordinates, because you will be blamed if anything goes wrong
   B. Be sure each subordinate understands the specific job he has been assigned, and check at intervals to make sure assignments are done properly
   C. Assign several people to every important job so that responsibility will be spread out as much as possible
   D. Have an experienced subordinate check all work done by other employees so that there will be little chance of anything going wrong

5. The human tendency to resist change is often reflected in higher rates of turnover, absenteeism, and errors whenever an important change is made in an organization. Although psychologists do not fully understand the reasons why people resist change, they believe that the resistance stems from a threat to the individual's security, that it is a form of fear of the unknown.
In light of this statement, which one of the following approaches would probably be MOST effective in preparing employees for a change in procedure in their unit?
   A. Avoid letting employees know anything about the change until the last possible moment
   B. Sympathize with employees who resent the change and let them know you share their doubts and fears
   C. Promise the employees that if the change turns out to be a poor one, you will allow them to suggest a return to the old system
   D. Make sure that employees know the reasons for the change and are aware of the benefits that are expected from it

6. Each of the following methods of encouraging employee participation in work planning has been used effectively with different kinds and sizes of employee groups.
Which one of the following methods would be MOST suitable for a group of four technically skilled employees?
   A. Discussions between the supervisor and a representative of the group
   B. A suggestion program with semi-annual awards for outstanding suggestions
   C. A group discussion summoned whenever a major problem remains unsolved for more than a month
   D. Day-to-day exchange of information, opinions, and experience

7. Of the following, the MOST important reason why a supervisor is given the authority to tell subordinates what work they should do, how they should do it, and when it should be done is that usually
   A. most people will not work unless there is someone with authority standing over them

B. work is accomplished more effectively if the supervisor plans and coordinates it
C. when division of work is left up to subordinates, there is constant arguing, and very little work is accomplished
D. subordinates are not familiar with the tasks to be performed

8. Fatigue is a factor that affects productivity in all work situations. However, a brief rest period will ordinarily serve to restore a person from fatigue.
According to this statement, which one of the following techniques is MOST likely to reduce the impact of fatigue on overall productivity in a unit?
   A. Scheduling several short breaks throughout the day
   B. Allowing employees to go home early
   C. Extending the lunch period an extra half hour
   D. Rotating job assignments every few weeks

8.____

9. After giving a new task to an employee, it is a good idea for a supervisor to ask specific questions to make sure that the employee grasps the essentials of the task and sees how it can be carried out. Questions which ask the employee what he thinks or how he feels about an important aspect of the task are particularly effective.
Which one of the following questions is NOT the type of question which would be useful in the foregoing situation?
   A. Do you feel there will be any trouble meeting the 4:30 deadline?
   B. How do you feel about the kind of work we do here?
   C. Do you think that combining those two steps will work all right?
   D. Can you think of any additional equipment you may need for this process?

9.____

10. Of the following, the LEAST important reason for having a *continuous* training program is that
    A. employees may forget procedures that they have already learned
    B. employees may develop shortcuts on the job that result in inaccurate work
    C. the job continue to change because of new procedures and equipment
    D. training is one means of measuring effectiveness and productivity on the job

10.____

11. In training a new employee, it is usually advisable to break down the job into meaningful parts and have the new employee master one part before going on to the next.
Of the following, the BEST reason for using this technique is to
    A. let the new employee know the reason for what he is doing and thus encourage him to remain in the unit
    B. make the employee aware of the importance of the work and encourage him to work harder
    C. show the employee that the work is easy so that he will be encouraged to work faster
    D. make it more likely that the employee will experience success and will be encouraged to continue learning the job

11.____

12. You may occasionally find a serious error in the work of one of your subordinates.
    Of the following, the BEST time to discuss such an error with an employee usually is
    A. immediately after the error is found
    B. after about two weeks, since you will also be able to point out some good things that the employee has accomplished
    C. when you have discovered a pattern of errors on the part of this employee so that he will not be able to dispute your criticism
    D. after the error results in a complaint by your own supervisor

    12.____

13. For very important announcements to the staff, a supervisor should usually use both written and oral communications. For example, when a new procedure is to be introduced, the supervisor can more easily obtain the group's acceptance by giving his subordinates a rough draft of the new procedure and calling a meeting of all his subordinates.
    The LEAST important benefit of this technique is that it will better enable the supervisor to
    A. explain why the change is necessary
    B. make adjustments in the new procedure to meet valid staff objections
    C. assign someone to carry out the new procedure
    D. answer questions about the new procedure

    13.____

14. Assume that, while you are interviewing an individual to obtain information, the individual pauses in the middle of an answer.
    The BEST of the following actions for you to take at that time is to
    A. correct any inaccuracies in what he has said
    B. remain silent until he continues
    C. explain your position on the matter being discussed
    D. explain that time is short and that he must complete his story quickly

    14.____

15. When you are interviewing someone to obtain information, the BEST of the following reasons for you to repeat certain of his exact words is to
    A. assure him that appropriate action will be taken
    B. encourage him to switch to another topic of discussion
    C. assure him that you agree with his point of view
    D. encourage him to elaborate on a point he has made

    15.____

16. Generally, when writing a letter, the use of precise words and concise sentences is
    A. *good*, because less time will be required to write the letter
    B. *bad*, because it is most likely that the reader will think the letter is unimportant and will not respond favorably
    C. *good*, because it is likely that your desired meaning will be conveyed to the reader
    D. *bad*, because your letter will be too brief to provide adequate information

    16.____

17. In which of the following cases would it be MOST desirable to have two cards for one individual in a single alphabetic file?
The individual has
    A. a hyphenated surname
    B. two middle names
    C. a first name with an unusual spelling
    D. a compound first name

17.____

18. Of the following, it is MOST appropriate to use a form letter when it is necessary to answer many
    A. requests or inquiries from a single individual
    B. follow-up letters from individuals requesting additional information
    C. request or inquiries about a single subject
    D. complaints from individuals that they have been unable to obtain various types of information

18.____

19. Assume that you are asked to make up a budget for your section for the coming year, and you are told that the most important function of the budget is its "control function."
Of the following, "control" in this context implies MOST NEARLY that
    A. you will probably be asked to justify expenditures in any category when it looks as though these expenditures are departing greatly from the amount budgeted
    B. your section will probably not be allowed to spend more than the budgeted amount in any given category, although it is always permissible to spend less
    C. your section will be required to spend the exact amount budgeted in every category
    D. the budget will be filed in the Office of the Comptroller so that when a year is over the actual expenditures can be compared with the amounts in the budget

19.____

20. In writing a report, the practice of taking up the LEAST important points *first* and the most important points *last* is a
    A. *good* technique, since the final points made in a report will make the greatest impression on the reader
    B. *good* technique, since the material is presented in a more logical manner and will lead directly to the conclusions
    C. *poor* technique, since the reader's time is wasted by having to review irrelevant information before finishing the report
    D. *poor* technique, since it may cause the reader to lose interest in the report and arrive at incorrect conclusions about the report

20.____

21. Typically, when the technique of "supervision by results" is practiced, higher management sets down, either implicitly or explicitly, certain performance standards or goals that the subordinate is expected to meet. So long as these standards are met, management interferes very little.
The MOST likely result of the use of this technique is that it will

21.____

A. lead to ambiguity in terms of goals
B. be successful only to the extent that close direct supervision is practiced
C. make it possible to evaluate both employee and supervisory effectiveness
D. allow for complete dependence on the subordinate's part

22. When making written evaluations and reviews of the performance of subordinates, it is usually ADVISABLE to
    A. avoid informing the employee of the evaluation if it is critical because it may create hard feelings
    B. avoid informing the employee of the evaluation whether critical or favorable because it is tension-producing
    C. to permit the employee to see the evaluation but not to discuss it with him because the supervisor cannot be certain where the discussion might lead
    D. to discuss the evaluation openly with the employee because it helps the employee understand what is expected of him

23. There are a number of well-known and respected human relations principles that successful supervisors have been using for years in building good relationships with their employees.
    Which of the following does NOT illustrate such a principle?
    A. Give clear and complete instructions
    B. Let each person know how he is getting along
    C. Keep an open-door policy
    D. Make all relationships personal ones

24. Assume that it is necessary for you to give an unpleasant assignment to one of your subordinates. You expect this employee to raise some objections to this assignment.
    The MOST appropriate of the following actions for you to take FIRST is to issue the assignment
    A. *orally*, with the further statement that you will not listen to any complaints
    B. *in writing*, to forestall any complaints by the employee
    C. *orally*, permitting the employee to express his feelings
    D. *in writing*, with a note that any comments should be submitted in writing

25. Suppose you have just announced at a staff meeting with your subordinates that a radical reorganization of work will take place next week. Your subordinates at the meeting appear to be excited, tense, and worried.
    Of the following, the BEST action for you to take at that time is to
    A. schedule private conferences with each subordinate to obtain his reaction to the meeting
    B. close the meeting and tell your subordinates to return immediately to their work assignments
    C. give your subordinates some time to ask questions and discuss your announcement
    D. insist that your subordinates do not discuss your announcement among themselves or with other members of the agency

KEY (CORRECT ANSWERS)

1.	C		11.	D
2.	A		12.	A
3.	C		13.	C
4.	B		14.	B
5.	D		15.	D
6.	D		16.	C
7.	B		17.	A
8.	A		18.	C
9.	B		19.	A
10.	D		20.	D

21. C
22. D
23. D
24. C
25. C

TEST 2

DIRECTIONS: Each question or incomplete statement is followed by several suggested answers or completions. Select the one that BEST answers the question or completes the statement. *PRINT THE LETTER OF THE CORRECT ANSWER IN THE SPACE AT THE RIGHT.*

1. Of the following, the BEST way for a supervisor to increase employees' interest in their work is to
   A. allow them to make as many decisions as possible
   B. demonstrate to them that he is as technically competent as they
   C. give each employee a difficult assignment
   D. promptly convey to them instructions from higher management

   1.____

2. The one of the following which is LEAST important in maintaining a high level of productivity on the part of employees is the
   A. provision of optimum physical working conditions for employees
   B. strength of employees' aspirations for promotion
   C. anticipated satisfactions which employees hope to derive from their work
   D. employees' interest in their jobs

   2.____

3. Of the following, the MAJOR advantage of group problem-solving, as compared to individual problem-solving, is that groups will more readily
   A. abide by their own decisions
   B. agree with agency management
   C. devise new policies and procedures
   D. reach conclusions sooner

   3.____

4. The group problem-solving conference is a useful supervisory method for getting people to reach solutions to problems.
   Of the following, the reason that groups usually reach more realistic solutions than do individuals is that
   A. individuals, as a rule, take longer than do groups in reaching decisions and are, therefore, more likely to make an error
   B. bringing people together to let them confer impresses participants with the seriousness of problems
   C. groups are generally more concerned with the future in evaluating organizational problems
   D. the erroneous opinions of group members tend to be corrected by the other members

   4.____

5. A competent supervisor should be able to distinguish between human and technical problems.
   Of the following, the MAJOR difference between such problems is that serious human problems, in comparison to ordinary technical problems
   A. are remedied more quickly
   B. involve a lesser need for diagnosis
   C. are more difficult to define
   D. become known through indications which are usually the actual problem

   5.____

2 (#2)

6. Of the following, the BEST justification for a public agency establishing an alcoholism program for its employees is that
   A. alcoholism has traditionally been looked upon with a certain amused tolerance by management and thereby ignored as a serious illness
   B. employees with drinking problems have twice as many on-the-job accidents, especially during the early years of the problem
   C. excessive use of alcohol is associated with personality instability hindering informal social relationships among peers and subordinates
   D. the agency's public reputation will suffer despite an employee's drinking problem being a personal matter of little public concern

7. Assume you are a manager and you find a group of maintenance employees assigned to your project drinking and playing cards for money in an incinerator room after their regular working hours.
   The one of the following actions it would be BEST for you to take is to
   A. suspend all employees immediately if there is no question in your mind as to the validity of the charges
   B. review the personnel records of those involved with the supervisor and make a joint decision on which employees should sustain penalties of loss of annual leave or fines
   C. ask the supervisor to interview each violator and submit written reports to you and thereafter consult with the supervisor about disciplinary actions
   D. deduct three days of annual leave from each employee involved if he pleads guilty in lieu of facing more serious charges

8. Assume that as a manager you must discipline a subordinate, but all of the pertinent facts necessary for a full determination of the appropriate action to take are not yet available. However, you fear that a delay in disciplinary action may damage the morale of other employees.
   The one of the following which is MOST appropriate for you to do in this matter is to
   A. take immediate disciplinary action as if all the pertinent facts were available
   B. wait until all pertinent facts are available before reaching a decision
   C. inform the subordinate that you know he is guilty, issue a stern warning, and then let him wait for your further action
   D. reduce the severity of the discipline appropriate for the violation

9. There are two standard dismissal procedures utilized by most public agencies. The first is the "open back door" policy, in which the decision of a supervisor in discharging an employee for reasons of inefficiency cannot be cancelled by the central personnel agency. The second is the "closed back door" policy, in which the central personnel agency can order the supervisor to restore the discharged employee to his position.
   Of the following, the major DISADVANTAGE of the "closed back door" policy as opposed to the "open back door" policy is that central personnel agencies are
   A. likely to approve the dismissal of employees when there is inadequate justification

B. likely to revoke dismissal actions out of sympathy for employees
C. less qualified than employing agencies to evaluate the efficiency of employees
D. easily influenced by political, religious, and racial factors

10. The one of the following for which a formal grievance-handling system is LEAST useful is in
    A. reducing the frequency of employee complaints
    B. diminishing the likelihood of arbitrary action by supervisors
    C. providing an outlet for employee frustrations
    D. bringing employee problems to the attention of higher management

11. The one of the following managers whose leadership style involves the GREATEST delegation of authority to subordinates is the one who presents to subordinates
    A. his ideas and invites questions
    B. his decision and persuades them to accept it
    C. the problem, gets their suggestions, and makes his decision
    D. a tentative decision which is subject to change

12. Which of the following is MOST likely to cause employee productivity standards to be set too high?
    A. Standards of productivity are set by first-line supervisors rather than by higher level managers.
    B. Employees' opinions about productivity standards are sought through written questionnaires.
    C. Initial studies concerning productivity are conducted by staff specialists.
    D. Ideal work conditions assumed in the productivity standards are lacking in actual operations.

13. The one of the following which states the MAIN value of an organization chart for a manager is that such charts show the
    A. lines of formal authority
    B. manner in which duties are performed by each employee
    C. flow of work among employees on the same level
    D. specific responsibilities of each position

14. Which of the following BEST names the usual role of a line unit with regard to the organization's programs?
    A. Seeking publicity   B. Developing
    C. Carrying out   D. Evaluating

15. Critics of promotion *from within* a public agency argue for hiring *from outside* the agency because they believe that promotion from within leads to
    A. resentment and consequent weakened morale on the part of those not promoted
    B. the perpetuation of outdated practices and policies
    C. a more complex hiring procedure than hiring from outside the agency
    D. problems of objectively appraising someone already in the organization

4 (#2)

16. The one of the following management functions which usually can be handled MOST effectively by a committee is the
    A. settlement of interdepartmental disputes
    B. planning of routine work schedules
    C. dissemination of information
    D. assignment of personnel

    16.____

17. Assume that you are serving on a committee which is considering proposals in order to recommend a new maintenance policy. After eliminating a number of proposals by unanimous consent, the committee is deadlocked on three proposals.
    The one of the following which is the BEST way for the committee to reach agreement on a proposal they could recommend is to
    A. consider and vote on each proposal separately by secret ballot
    B. examine and discuss the three proposals until the proponents of two of them are persuaded they are wrong
    C. reach a synthesis which incorporates the significant features of each proposals
    D. discuss the three proposals until the proponents of each one concede those aspects of the proposals about which there is disagreement

    17.____

18. A commonly used training and development method for professional staff is the case method, which utilizes the description of a situation, real or simulated, to provide a common base for analysis, discussion, and problem-solving.
    Of the following, the MOST appropriate time to use the case method is when professional staff needs
    A. insight into their personality problems
    B. practice in applying management concepts to their own problems
    C. practical experience in the assignment of delegated responsibilities
    D. to know how to function in many different capacities

    18.____

19. The incident process is a training and development method in which trainees are given a very brief statement of an event or o a situation presenting a job incident or an employee problem of special significance.
    Of the following, it is MOST appropriate to use the incident process when
    A. trainees need to learn to review and analyze facts before solving a problem
    B. there are a large number of trainees who require the same information
    C. there are too many trainees to carry on effective discussion
    D. trainees are not aware of the effect of their behavior on others

    19.____

20. The one of the following types of information about which a clerical employee is usually LEAST concerned during the orientation process is
    A. his specific job duties   B. where he will work
    C. his organization's history   D. who his associates will be

    20.____

21. The one of the following which is the MOST important limitation on the degree to which work should be broken down into specialized tasks is the point at which
    A. there ceases to be sufficient work of a specialized nature to occupy employees
    B. training costs equal the half-yearly savings derived from further specialization
    C. supervision of employees performing specialized tasks becomes more technical than supervision of general employees
    D. it becomes more difficult to replace the specialist than to replace the generalist who performs a complex set of functions

21.____

22. When a supervisor is asked for his opinion of the suitability for promotion of a subordinate, the supervisor is actually being asked to predict the subordinate's future behavior in a new role.
    Such a prediction is MOST likely to be accurate if the
    A. higher position is similar to the subordinate's current one
    B. higher position requires intangible personal qualities
    C. new position has had little personal association with the subordinate away from the job

22.____

23. In one form of the non-directive evaluation interview, the supervisor communicates his evaluation to the employee and then listens to the employee's response without making further suggestions.
    The one of the following which is the PRINCIPAL danger of this method of evaluation is that the employee is MOST likely to
    A. develop an indifferent attitude towards the supervisor
    B. fail to discover ways of improving his performance
    C. become resistant to change in the organization's structure
    D. place the blame for his shortcomings on his co-workers

23.____

24. In establishing rules for his subordinates, a superior should be PRIMARILY concerned with
    A. creating sufficient flexibility to allow for exceptions
    B. making employees aware of the reasons for the rules and the penalties for infractions
    C. establishing the strength of his own position in relation to his subordinates
    D. having his subordinates know that such rules will be imposed in a personal manner

24.____

25. The practice of conducting staff training sessions on a periodic basis is generally considered
    A. *poor*; it takes employees away from their work assignments
    B. *poor*; all staff training should be done on an individual basis
    C. *good*; it permits the regular introduction of new methods and techniques
    D. *good*; it ensures a high employee productivity rate

25.____

KEY (CORRECT ANSWERS)

1.	A		11.	C
2.	A		12.	D
3.	A		13.	A
4.	D		14.	C
5.	C		15.	B
6.	B		16.	A
7.	C		17.	C
8.	B		18.	B
9.	C		19.	A
10.	A		20.	C

21.	A
22.	A
23.	B
24.	B
25.	C

PREPARING WRITTEN MATERIALS
EXAMINATION SECTION
TEST 1

DIRECTIONS: Each question consists of a sentence which may be classified appropriately under one of the following four categories:
- A. Incorrect because of faulty grammar or sentence structure.
- B. Incorrect because of faulty punctuation.
- C. Incorrect because of faulty spelling or capitalization.
- D. Correct

Examine each sentence carefully. Then, in the space at the right, print the capital letter preceding the option which is the BEST of the four suggested above. All incorrect sentences contain only one type of error. Consider a sentence correct if it contains none of the types of errors mentioned, although there may be other correct ways of expressing the same thought.

1. The fire apparently started in the storeroom, which is usually locked. 1.____
2. On approaching the victim two bruises were noticed by this officer. 2.____
3. The officer, who was there examined the report with great care. 3.____
4. Each employee in the office had a separate desk. 4.____
5. The suggested procedure is similar to the one now in use. 5.____
6. No one was more pleased with the new procedure than the chauffeur. 6.____
7. He tried to pursuade her to change the procedure. 7.____
8. The total of the expenses charged to petty cash were high. 8.____
9. An understanding between him and I was finally reached. 9.____
10. It was at the supervisor's request that the clerk agreed to postpone his vacation. 10.____
11. We do not believe that it is necessary for both he and the clerk to attend the conference. 11.____
12. All employees, who display perseverance, will be given adequate recognition. 12.____
13. He regrets that some of us employees are dissatisfied with our new assignments. 13.____

14. "Do you think that the raise was merited," asked the supervisor?   14.____

15. The new manual of procedure is a valuable supplament to our rules and regulation.   15.____

16. The typist admitted that she had attempted to pursuade the other employees to assist her in her work.   16.____

17. The supervisor asked that all amendments to the regulations be handled by you and I.   17.____

18. They told both he and I that the prisoner had escaped.   18.____

19. Any superior officer, who, disregards the just complaints of his subordinates, is remiss in the performance of his duty.   19.____

20. Only those members of the national organization who resided in the Middle west attended the conference in Chicago.   20.____

21. We told him to give the investigation assignment to whoever was available.   21.____

22. Please do not disappoint and embarass us by not appearing in court.   22.____

23. Despite the efforts of the Supervising mechanic, the elevator could not be started.   23.____

24. The U.S. Weather Bureau, weather record for the accident date was checked.   24.____

KEY (CORRECT ANSWERS)

1.	D		11.	A
2.	A		12.	B
3.	B		13.	D
4.	D		14.	B
5.	D		15.	C
6.	D		16.	C
7.	C		17.	A
8.	A		18.	A
9.	A		19.	B
10.	D		20.	C

21. D
22. C
23. C
24. B

TEST 2

DIRECTIONS: Each question consists of a sentence. Some of the sentences contain errors in English grammar or usage, punctuation, spelling, or capitalization. A sentence does not contain an error simply because it could be written in a different manner. Choose answer:
- A. If the sentence contains an error in English grammar or usage.
- B. if the sentence contains an error in punctuation.
- C. If the sentence contains an error in spelling or capitalization
- D. If the sentence does not contain any errors.

1. The severity of the sentence prescribed by contemporary statutes—including both the former and the revised New York Penal Laws—do not depend on what crime was intended by the offender.

2. It is generally recognized that two defects in the early law of attempt played a part in the birth of burglary: (1) immunity from prosecution for conduct short of the last act before completion of the crime, and (2) the relatively minor penalty imposed for an attempt (it being a common law misdemeanor) vis-à-vis the completed offense.

3. The first sentence of the statute is applicable to employees who enter their place of employment, invited guests, and all other persons who have an express or implied license or privilege to enter the premises.

4. Contemporary criminal codes in the United States generally divide burglary into various degrees, differentiating the categories according to place, time and other attendent circumstances.

5. The assignment was completed in record time but the payroll for it has not yet been prepaid.

6. The operator, on the other hand, is willing to learn me how to use the mimeograph.

7. She is the prettiest of the three sisters.

8. She doesn't know; if the mail has arrived.

9. The doorknob of the office door is broke.

10. Although the department's supply of scratch pads and stationery have diminished considerably, the allotment for our division has not been reduced.

11. You have not told us whom you wish to designate as your secretary.

12. Upon reading the minutes of the last meeting, the new proposal was taken up for consideration.

2 (#2)

13. Before beginning the discussion, we locked the door as a precautionery measure.   13._____

14. The supervisor remarked, "Only those clerks, who perform routine work, are permitted to take a rest period."   14._____

15. Not only will this duplicating machine make accurate copies, but it will also produce a quantity of work equal to fifteen transcribing typists.   15._____

16. "Mr. Jones," said the supervisor, "we regret our inability to grant you an extention of your leave of absence."   16._____

17. Although the employees find the work monotonous and fatigueing, they rarely complain.   17._____

18. We completed the tabulation of the receipts on time despite the fact that Miss Smith our fastest operator was absent for over a week.   18._____

19. The reaction of the employees who attended the meeting, as well as the reaction of those who did not attend, indicates clearly that the schedule is satisfactory to everyone concerned.   19._____

20. Of the two employees, the one in our office is the most efficient.   20._____

21. No one can apply or even understand, the new rules and regulations.   21._____

22. A large amount of supplies were stored in the empty office.   22._____

23. If an employee is occassionally asked to work overtime, he should do so willingly.   23._____

24. It is true that the new procedures are difficult to use but, we are certain that you will learn them quickly.   24._____

25. The office manager said that he did not know who would be given a large allotment under the new plan.   25._____

KEY (CORRECT ANSWERS)

1.	A	11.	D
2.	D	12.	A
3.	D	13.	C
4.	C	14.	B
5.	C	15.	A
6.	A	16.	C
7.	D	17.	C
8.	B	18.	B
9.	A	19.	D
10.	A	20.	A

21. B
22. A
23. C
24. B
25. D

TEST 3

DIRECTIONS: Each of the following sentences may be classified MOST appropriately under one of the following categories:
  A. Faulty because of incorrect grammar
  B. Faulty because of incorrect punctuation
  C. Faulty because of incorrect capitalization
  D. Correct

Examine each sentence carefully. Then, in the space at the right, print the capital letter preceding the option which is the BEST of the four suggested above. All incorrect sentence contain but one type of error. Consider a sentence correct if it contains none of the types of errors mentioned, even though there may be other correct ways of expressing the same thought.

1. The desk, as well as the chairs, were moved out of the office. 1.____

2. The clerk whose production was greatest for the month won a day's vacation as first prize. 2.____

3. Upon entering the room, the employees were found hard at work at their desks. 3.____

4. John Smith our new employee always arrives at work on time. 4.____

5. Punish whoever is guilty of stealing the money. 5.____

6. Intelligent and persistent effort lead to success no matter what the job may be. 6.____

7. The secretary asked, "can you call again at three o'clock?" 7.____

8. He told us, that if the report was not accepted at the next meeting, it would have to be rewritten. 8.____

9. He would not have sent the letter if he had known that it would cause so much excitement. 9.____

10. We all looked forward to him coming to visit us. 10.____

11. If you find that you are unable to complete the assignment please notify me as soon as possible. 11.____

12. Every girl in the office went home on time but me; there was still some work for me to finish. 12.____

13. He wanted to know who the letter was addressed to, Mr. Brown or Mr. Smith. 13.____

14. "Mr. Jones, he said, please answer this letter as soon as possible." 14.____

15. The new clerk had an unusual accent inasmuch as he was born and educated in the south.   15.____

16. Although he is younger than her, he earns a higher salary.   16.____

17. Neither of the two administrators are going to attend the conference being held in Washington, D.C.   17.____

18. Since Miss Smith and Miss Jones have more experience than us, they have been given more responsible duties.   18.____

19. Mr. Shaw the supervisor of the stock room maintains an inventory of stationery and office supplies.   19.____

20. Inasmuch as this matter affects both you and I, we should take joint action.   20.____

21. Who do you think will be able to perform this highly technical work?   21.____

22. Of the two employees, John is considered the most competent.   22.____

23. He is not coming home on tuesday; we expect him next week.   23.____

24. Stenographers, as well as typists must be able to type rapidly and accurately.   24.____

25. Having been placed in the safe we were sure that the money would not be stolen.   25.____

KEY (CORRECT ANSWERS)

1.	A		11.	B
2.	D		12.	D
3.	A		13.	A
4.	B		14.	B
5.	D		15.	C
6.	A		16.	A
7.	C		17.	A
8.	B		18.	A
9.	D		19.	B
10.	A		20.	A

21.	D
22.	A
23.	C
24.	B
25.	A

TEST 4

DIRECTIONS: Each of the following sentences consist of four sentences lettered A, B, C, and D. One of the sentences in each group contains an error in grammar or punctuation. Indicate the INCORRECT sentence in each group. *PRINT THE LETTER OF THE CORRECT ANSWER IN THE SPACE AT THE RIGHT.*

1. A. Give the message to whoever is on duty.
   B. The teacher who's pupil won first prize presented the award.
   C. Between you and me, I don't expect the program to succeed.
   D. His running to catch the bus caused the accident.

   1.____

2. A. The process, which was patented only last year is already obsolete.
   B. His interest in science (which continues to the present) led him to convert his basement into a laboratory.
   C. He described the book as "verbose, repetitious, and bombastic".
   D. Our new director will need to possess three qualities: vision, patience, and fortitude.

   2.____

3. A. The length of ladder trucks varies considerably.
   B. The probationary fireman reported to the officer to who he was assigned.
   C. The lecturer emphasized the need for we firemen to be punctual.
   D. Neither the officers nor the members of the company knew about the new procedure.

   3.____

4. A. Ham and eggs is the specialty of the house.
   B. He is one of the students who are on probation.
   C. Do you think that either one of us have a chance to be nominated for president of the class?
   D. I assume that either he was to be in charge or you were.

   4.____

5. A. Its a long road that has no turn.
   B. To run is more tiring than to walk.
   C. We have been assigned three new reports: namely, the statistical summary, the narrative summary, and the budgetary summary.
   D. Had the first payment been made in January, the second would be due in April.

   5.____

6. A. Each employer has his own responsibilities.
   B. If a person speaks correctly, they make a good impression.
   C. Every one of the operators has had her vacation.
   D. Has anybody filed his report?

   6.____

7. A. The manager, with all his salesmen, was obliged to go.
   B. Who besides them is to sign the agreement?
   C. One report without the others is incomplete.
   D. Several clerks, as well as the proprietor, was injured.

   7.____

2 (#4)

8.
- A. A suspension of these activities is expected.
- B. The machine is economical because first cost and upkeep are low.
- C. A knowledge of stenography and filing are required for this position.
- D. The condition in which the goods were received shows that the packing was not done properly.

8.____

9.
- A. There seems to be a great many reasons for disagreement.
- B. It does not seem possible that they could have failed.
- C. Have there always been too few applicants for these positions?
- D. There is no excuse for these errors.

9.____

10.
- A. We shall be pleased to answer your question.
- B. Shall we plan the meeting for Saturday?
- C. I will call you promptly at seven.
- D. Can I borrow your book after you have read it?

10.____

11.
- A. You are as capable as I.
- B. Everyone is willing to sign but him and me.
- C. As for he and his assistant, I cannot praise them too highly.
- D. Between you and me, I think he will be dismissed.

11.____

12.
- A. Our competitors bid above us last week.
- B. The survey which was began last year has not yet been completed.
- C. The operators had shown that they understood their instructions.
- D. We have never ridden over worse roads.

12.____

13.
- A. Who did they say was responsible?
- B. Whom did you suspect?
- C. Who do you suppose it was?
- D. Whom do you mean?

13.____

14.
- A. Of the two propositions, this is the worse.
- B. Which report do you consider the best—the one in January or the one in July?
- C. I believe this is the most practicable of the many plans submitted.
- D. He is the youngest employee in the organization.

14.____

15.
- A. The firm had but three orders last week.
- B. That doesn't really seem possible.
- C. After twenty years scarcely none of the old business remains.
- D. Has he done nothing about it?

15.____

KEY (CORRECT ANSWERS)

1.	B	6.	B	11.	C
2.	A	7.	D	12.	B
3.	C	8.	C	13.	A
4.	C	9.	A	14.	B
5.	A	10.	D	15.	C

PREPARING WRITTEN MATERIAL

PARAGRAPH REARRANGEMENT
COMMENTARY

The sentences that follow are in scrambled order. You are to rearrange them in proper order and indicate the letter choice containing the correct answer at the space at the right.

Each group of sentences in this section is actually a paragraph presented in scrambled order. Each sentence in the group has a place in that paragraph; no sentence is to be left out. You are to read each group of sentences and decide upon the best order in which to put the sentences so as to form a well-organized paragraph.

The questions in this section measure the ability to solve a problem when all the facts relevant to its solution are not given.

More specifically, certain positions of responsibility and authority require the employee to discover connection between events sometimes, apparently, unrelated. In order to do this, the employee will find it necessary to correctly infer that unspecified events have probably occurred or are likely to occur. This ability becomes especially important when action must be taken on incomplete information.

Accordingly, these questions require competitors to choose among several suggested alternatives, each of which presents a different sequential arrangement of the events. Competitors must choose the MOST logical of the suggested sequences.

In order to do so, they may be required to draw on general knowledge to infer missing concepts or events that are essential to sequencing the given events. Competitors should be careful to infer only what is essential to the sequence. The plausibility of the wrong alternatives will always require the inclusion of unlikely events or of additional chains of events which are NOT essential to sequencing the given events.

It's very important to remember that you are looking for the best of the four possible choices, and that the best choice of all may not even be one of the answers you're given to choose from.

There is no one right way to solve these problems. Many people have found it helpful to first write out the order of the sentences, as they would have arranged them, on their scrap paper before looking at the possible answers. If their optimum answer is there, this can save them some time. If it isn't, this method can still give insight into solving the problem. Others find it most helpful to just go through each of the possible choices, contrasting each as they go along. You should use whatever method feels comfortable and works for you.

While most of these types of questions are not that difficult, we've added a higher percentage of the difficult type, just to give you more practice. Usually there are only one or two questions on this section that contain such subtle distinctions that you're unable to answer confidently. And you then may find yourself stuck deciding between two possible choices, neither of which you're sure about.

EXAMINATION SECTION
TEST 1

DIRECTIONS: The sentences that follow are in scrambled order. You are to rearrange them in proper order and indicate the letter choice containing the CORRECT answer. *PRINT THE LETTER OF THE CORRECT ANSWER IN THE SPACE AT THE RIGHT.*

1. Fire Marshal Adams has arrested a man for pulling a false alarm. He has recorded the following items of information about the incident in his notebook for use in his subsequent report:
    I. I was on surveillance at a frequently pulled false alarm box located at Edison Street and Harvard Road.
    II. At 1605 hours, I observed the white male, with long brown hair and a mustache, wearing black pants and a red shirt, pull the fire alarm box.
    III. I interviewed the officer of the first due ladder company, Lt. Morgan - L-37, who informed me that a search of the area disclosed no cause for an alarm to be transmitted.
    IV. A man wearing a red shirt, black pants, with long brown hair and a mustache came out of Ryan's Pub, located at Edison Street and Harvard Road, and walked directly to the alarm box.
    V. I stopped the man about five blocks away at 33rd Street and Harvard Road and asked him why he pulled the fire alarm box, and he replied, *Because I felt like it.*

    The MOST logical order for the above sentences to appear in the report is

    A. I, IV, II, III, V
    B. I, II, III, IV, V
    C. I, IV, III, II, V
    D. I, IV, V, II, III

    1.____

2. A fire marshal is preparing a report regarding Tom Jones, who was a witness to an arson fire at his apartment building. Following are five sentences which will be included in the report:
    I. On July 16, I responded to the fire building, address 2020 Elm Street, to interview Tom Jones.
    II. Tom Jones described the *super* (name unknown) as a middle-aged male with beard, six feet tall, wearing a blue jumpsuit.
    III. Tom Jones stated that he saw the *super* of the building next door set the fire.
    IV. After being advised of his constitutional rights at the 44th Precinct detective's squad room, the *super* confessed.
    V. I interviewed the *super* and took him to the precinct for further investigation.

    The MOST logical order for the above sentences to appear in the report is

    A. I, II, III, V, IV
    B. I, II, III, IV, V
    C. I, III, II, IV, V
    D. I, III, II, V, IV

    2.____

3. A fire marshal is preparing a report on a shooting incident which will include the following five sentences:
   I. I ran around the corner and observed a man pointing a gun at another man.
   II. I informed the man I was a police officer and that he should drop his gun.
   III. I was on the corner of 4th Avenue and 43rd Street when I heard a gunshot coming from around the corner.
   IV. The man turned around and pointed his gun at me.
   V. I fired once, shooting him in the chest and causing him to fall to the ground.
   The MOST logical order for the above sentences to appear in the report is

   A. I, III, IV, II, V
   B. IV, V, II, I, III
   C. III, I, II, IV, V
   D. III, I, V, II, IV

4. Fire Marshal Smith is writing a report. The report will include the following five sentences:
   I. I asked the woman for a description of the man and his location in the building.
   II. When I said, *Don't move, Five Marshal*, the man dropped the can containing a flammable liquid.
   III. I transmitted on my handie-talkie for fire companies to respond.
   IV. A woman approached our car and said there was a man pouring a liquid, which she thought to be gasoline, on a staircase at 123 East Street.
   V. Upon entering that location, I observed a man spilling a liquid on the floor.
   The MOST logical order for the above sentences to appear on the interview sheet is

   A. IV, I, V, II, III
   B. I, IV, III, V, II
   C. V, II, IV, I, III
   D. IV, III, I, V, II

5. Fire Marshal Fox is completing an interview report for a fire in the kitchen of an apartment at 1700 Clayton Road. The following five sentences will be included in the interview report:
   I. This is the first fire in which Mrs. Brown has ever been involved.
   II. A neighbor smelled the food burning and called the Fire Department.
   III. Mrs. Brown has been a tenant in Apt. 4C for 7 years.
   IV. Mrs. Brown was very tired and laid down to rest and fell asleep.
   V. Mrs. Brown was cooking beef stew in the kitchen after coming home from work.
   The MOST logical order for the above sentences to appear in the report is

   A. II, III, I, IV, V
   B. III, V, IV, II, I
   C. I, III, II, V, IV
   D. III, V, I, IV, II

6. A fire marshal is completing a report of an arson fire. The report will contain the following five statements made by a witness:
   I. I heard the sound of breaking glass; and when I looked out my window, I saw orange flames coming from the building across the street.
   II. I saw two young men on bicycles rapidly riding away, one with long blond hair, the other had long brown hair.
   III. He made a threat to get even when he was being evicted.
   IV. The young man with long blond hair was evicted from the fire building last week.
   V. The two young men rode in the direction of Flowers Avenue.
   The MOST logical order for the above statements to appear in the report is

A. I, II, V, IV, III   B. I, II, IV, V, III
C. III, I, V, II, IV   D. III, I, II, IV, V

7. A fire marshal is preparing a report regarding an eleven-year-old who was burned in a fire at the Midtown School for Boys. The report will include the following five sentences:
    I. The child described the fire-setter as a male with glasses, five feet tall, wearing a blue uniform.
    II. On December 12, I responded to Hill Top Hospital to interview a child who was burned in a fire at the Midtown School for Boys.
    III. The male perpetrator made a full confession in front of the Assistant District Attorney at the precinct.
    IV. I responded to the school, after interviewing the boy, and found a security guard who fit the description.
    V. I interviewed the security guard and took him to the precinct for further questioning.

    The MOST logical order for the above sentences to appear in the fire report is

    A. I, IV, V, II, III   B. IV, III, II, I, V
    C. II, I, IV, V, III   D. II, IV, I, V, III

8. A fire marshal is preparing a report concerning a fire in an auto body shop. The report will contain the following five sentences:
    I. The shop owner stated that he argued with a customer about the cost of a repair job.
    II. The shop owner will be the complainant in the arson case.
    III. While on surveillance, my partner and I saw the fire and called it in over the Department radio.
    IV. The customer paid the bill and left saying, *I'll fix you for charging so much.*
    V. According to witnesses, the customer returned to the shop and threw a Molotov cocktail on the floor.

    The MOST logical order for the above sentences to appear in the report is

    A. I, IV, V, II, III   B. III, I, IV, V, II
    C. V, I, IV, III, II   D. III, V, I, IV, II

9. Security Officer Mace is completing an entry in her memo-book. The entry has the following five sentences:
    I. I observed the defendant removing a radio from a facility vehicle.
    II. I placed the defendant under arrest and escorted him to the patrolroom.
    III. I was patrolling the facility parking lot.
    IV. I asked the defendant to show identification. V. I determined that the defendant was not authorized to remove the radio.

    The MOST logical order for these sentences to be entered in Officer Mace's memo-book is

    A. I, III, II, IV, V   B. II, V, IV, I, III
    C. III, I, IV, V, II   D. IV, V, II, I, III

10. Security Officer Riley is completing an entry in his memo-book. The entry has the following five sentences:
    I. Anna Jones admitted that she stole Mary Green's wallet.
    II. I approached the women and asked them who they were and why they were arguing.
    III. I arrested Anna Jones for stealing Mary Green's wallet.
    IV. They identified themselves and Mary Green accused Anna Jones of stealing her wallet.
    V. I was in the lobby area when I observed two women arguing about a wallet.

    The MOST logical order for these sentences to be entered in Officer Riley's memo-book is

    A. II, IV, I, III, V
    B. III, I, IV, V, II
    C. IV, I, V, II, III
    D. V, II, IV, I, III

11. Assume that Security Officer John Ryan is completing an entry in his memobook. The entry has the following five sentences:
    I. I then cleared the immediate area of visitors and staff.
    II. I noticed smoke coming from a broom closet outside Room A71.
    III. Sergeant Mueller arrived with other officers to assist in clearing the area.
    IV. Upon investigation, I determined the smoke was due to burning material in the broom closet.
    V. I pulled the corridor fire alarm and notified Sergeant Mueller of the fire.

    The MOST logical order for these sentences to be entered in Officer Ryan's memo-book is

    A. II, III, IV, V, I
    B. II, IV, V, I, III
    C. IV, I, II, III, V
    D. V, III, II, I, IV

12. Security Officer Hernandez is completing an entry in his memobook. The entry has the following five sentences:
    I. I asked him to leave the premises immediately.
    II. A visitor complained that there was a strange man loitering in Clinic B hallway.
    III. I went to investigate and saw a man dressed in rags sitting on the floor of the hallway.
    IV. As he walked out, he started yelling that he had no place to go.
    V. I asked to see identification, but he said that he did not have any.

    The MOST logical order for these sentences to be entered in Officer Hernandez's memobook is

    A. II, III, V, I, IV
    B. III, I, II, IV, V
    C. IV, I, V, II, III
    D. III, I, V, II, IV

13. Officer Hogan is completing an entry in his memobook. The entry has the following five sentences:
    I. When the fighting had stopped, I transmitted a message requesting medical assistance for Mr. Perkins.
    II. Special Officer Manning assisted me in stopping the fight,
    III. When I arrived at the scene, I saw a client, Adam Finley, strike a facility employee, Peter Perkins.
    IV. As I attempted to break up the fight, Special Officer Manning came on the scene.
    V. I received a radio message from Sergeant Valez to investigate a possible fight in progress in the waiting room.

    The MOST logical order for these sentences to be entered in Officer Hogan's memo-book is

    A. II, I, IV, V, III
    B. III, V, II, IV, I
    C. IV, V, III, I, II
    D. V, III, IV, II, I

14. Police Officer White is preparing a crime report concerning the burglary of Mr. Smith's home. The report will contain the following five sentences:
    I. Upon entering the house, Mr. Smith noticed that the mortgage money, which had been left on the kitchen table, had been taken.
    II. An investigation by the reporting Officer determined that the burglar had left the house through the first floor rear door.
    III. Further investigation revealed that there were no witnesses to the burglary.
    IV. In addition, several pieces of jewelry were missing from a first floor bedroom.
    V. After arriving at home, Mr. Smith discovered that someone had broken into the house by jimmying the front door.

    The MOST logical order for the above sentences to appear in the report is

    A. V, IV, II, III, I
    B. V, I, III, IV, II
    C. V, I, IV, II, III
    D. V, IV, II, I, III

15. Police Officer Jenner responds to the scene of a burglary at 2106 La Vista Boulevard. He is approached by an elderly man named Richard Jenkins, whose account of the incident includes the following five sentences:
    I. I saw that the lock on my apartment door had been smashed and the door was open.
    II. My apartment was a shambles; my belongings were everywhere and my television set was missing.
    III. As I walked down the hallway toward the bedroom, I heard someone opening a window.
    IV. I left work at 5:30 P.M. and took the bus home.
    V. At that time, I called the police.

    The MOST logical order for the above sentences to appear in the report is

    A. I, V, IV, II, III
    B. IV, I, II, III, V
    C. I, V, II, III, IV
    D. IV, III, II, V, I

16. Police Officer LaJolla is writing an Incident Report in which back-up assistance was required. The report will contain the following five sentences:
    I. The radio dispatcher asked what my location was and he then dispatched patrol cars for back-up assistance.
    II. At approximately 9:30 P.M., while I was walking my assigned footpost, a gunman fired three shots at me.
    III. I quickly turned around and saw a White male, approximately 5'10", with black hair, wearing blue jeans, a yellow T-shirt, and white sneakers, running across the avenue carrying a handgun.
    IV. When the back-up officers arrived, we searched the area but could not find the suspect.
    V. I advised the radio dispatcher that a gunman had just fired a gun at me, and then I gave the dispatcher a description of the man.
    The MOST logical order for the above sentences to appear in the report is

    A. III, V, II, IV, I
    B. II, III, V, I, IV
    C. III, II, IV, I, V
    D. II, V, I, III, IV

17. Police Officer Engle is completing a Complaint Report of a burglary which occurred at Monty's Bar. The following five sentences will be included in the Complaint Report:
    I. The owner said that approximately $600 was taken, along with eight bottles of expensive brandy.
    II. The burglar apparently gained entry to the bar through the window and exited through the front door.
    III. When Mr. Barrett returned to reopen the bar at 1:00 P.M., he found the front door open and items thrown all over the bar.
    IV. Mr. Barrett, the owner of Monty's Bar, said he closed the bar at 4:00 M. and locked all the doors.
    V. After interviewing the owner, I conducted a search of the bar and found that a window in the back of the bar was broken.
    The MOST logical order for the above sentences to appear in the report is

    A. II, IV, III, V, I
    B. IV, III, I, V, II
    C. IV, II, III, I, V
    D. II, V, IV, III, I

18. Police Officer Revson is writing a report concerning a vehicle pursuit. His report will include the following five sentences:
    I. I followed the vehicle for several blocks and then motioned to the driver to pull the car over to the curb and stop.
    II. I informed the radio dispatcher that I was in a high-speed pursuit.
    III. When the driver ignored me, I turned on my siren and the driver increased his speed.
    IV. The vehicle hit a tree, and I was able to arrest the driver.
    V. While on patrol in Car #4135, I observed a motorist driving suspiciously.
    The MOST logical order for the above sentences to appear in the report is

    A. V, I, III, II, IV
    B. II, V, III, I, IV
    C. V, I, II, IV, III
    D. II, I, V, IV, III

19. Crime Reports are completed by Police Officers. One section of a report contains the following five sentences:
    I. The man, seeing that the woman had the watch, pushed Mr. Lugano to the ground.
    II. Frank Lugano was walking into the Flame Diner on Queens Boulevard when he was jostled by a man in front of him.
    III. A few minutes later, Mr. Lugano told a police officer on foot patrol about a man and a woman taking his watch.
    IV. As soon as he was jostled, a woman reached toward Mr. Lugano's wrist and removed his expensive watch.
    V. The man and woman, after taking Mr. Lugano's watch, ran around the corner.

    The MOST logical order for the above sentences to appear in the report is

    A. II, IV, I, III, V
    B. II, IV, I, V, III
    C. IV, I, III, II, V
    D. IV, II, I, V, III

20. Detective Adams completed a Crime Report which includes the following five sentences:
    I. I arrived at the scene of the crime at 10:20 A.M. and began to question Mr. Sands about the security devices he had installed.
    II. Several clearly identifiable fingerprints were found.
    III. A Fingerprint Unit specialist arrived at the scene and immediately began to dust for fingerprints.
    IV. After questioning Mr. Sands, I called the Fingerprint Unit.
    V. On Friday morning at 10 A.M., Mr. Sands, the owner of the High Fashion Fur Store on Fifth Avenue, called the precinct to report that his safe had been broken into.

    The MOST logical order for the above sentences to appear in the Crime Report is

    A. I, V, IV, III, II
    B. I, V, III, IV, II
    C. V, I, IV, II, III
    D. V, I, IV, III, II

KEY (CORRECT ANSWERS)

1. A
2. D
3. C
4. A
5. B

6. A
7. C
8. B
9. C
10. D

11. B
12. A
13. D
14. C
15. B

16. B
17. B
18. A
19. B
20. D

TEST 2

DIRECTIONS: The sentences that follow are in scrambled order. You are to rearrange them in proper order and indicate the letter choice containing the CORRECT answer. *PRINT THE LETTER OF THE CORRECT ANSWER IN THE SPACE AT THE RIGHT.*

1. Police Officer Ling is preparing a Complaint Report of a missing person. His report will contain the following five sentences: 1.____
    - I. I was greeted by Mrs. Miah Ali, who stated her daughter Lisa, age 17, did not return from school.
    - II. I questioned Mrs. Ali as to what time her daughter left for school and what type of clothing she was wearing.
    - III. I notified the Patrol Sergeant, searched the building and area, and prepared a Missing Person Complaint Report.
    - IV. I received a call from the radio dispatcher to respond to 9 Maple Street, Apartment 1H, on a missing person complaint.
    - V. Mrs. Ali informed me that Lisa was wearing a grey suit and black shoes, and departed for school at 7:30 A.M.

    The MOST logical order for the above sentences to appear in the report is

    A. IV, I, V, II, III   B. I, IV, V, III, II
    C. IV, I, II, V, III   D. III, I, IV, II, V

2. Police Officer Dunn is preparing a Complaint Report which will include the following five sentences: 2.____
    - I. Mrs. Field screamed and fought with the man.
    - II. A man wearing a blue ski mask grabbed Mrs. Field's purse.
    - III. Mrs. Field was shopping on 34th Street and Broadway at 1 o'clock in the afternoon.
    - IV. The man then ran around the corner.
    - V. The man was white, five feet six inches tall with a medium build.

    The MOST logical order for the above sentences to appear in the report is

    A. I, V, II, IV, III   B. III, II, I, IV, V
    C. III, IV, V, I, II   D. V, IV, III, I, II

3. Police Officer Davis is preparing a written report concerning child abuse. The report will include the following five sentences: 3.____
    - I. I responded to the scene and was met by an adult and a child who was approximately four years old.
    - II. I was notified by an unidentified pedestrian of a possible case of child abuse at 325 Belair Terrace.
    - III. The adult told me that the child fell and that the police were not needed.
    - IV. I felt that this might be a case of child abuse, and I requested that a Sergeant respond to the scene.
    - V. The child was bleeding from the head and had several bruises on the face.

    The MOST logical order for the above sentences to appear in the report is

    A. II, I, V, III, IV   B. I, II, IV, III, V
    C. I, III, IV, II, V   D. II, IV, I, V, III

4. The following five sentences will be part of a memobook entry concerning found property:
   I. Mr. Gustav said that while cleaning the lobby he found six credit cards and a passport.
   II. The credit cards and passport were issued to Manuel Gomez.
   III. I went to the precinct to give the property to the Desk Officer.
   IV. I prepared a receipt listing the property, gave the receipt to Mr. Gustav, and had him sign my memobook.
   V. While on foot patrol, I was approached by Mr. Gustav, the superintendent of 50-12 Maiden Parkway.

   The MOST logical order for the above sentences to appear in the memobook is

   A. V, I, II, IV, III
   B. I, II, IV, III, V
   C. V, I, III, IV, II
   D. I, IV, III, II, V

5. Police Officer Thomas is making a memobook entry that will include the following five sentences:
   I. My partner obtained a brief description of the suspects and the direction they were heading when they left the store.
   II. Edward Lemkin was asked to come with us to search the immediate area.
   III. I transmitted this information over the radio.
   IV. At the corner of 72nd Street and Broadway, our patrol car was stopped by Edward Lemkin, the owner of PJ Records.
   V. He told us that a group of teenagers stole some merchandise from his record store.

   The MOST logical order for the above sentences to appear in the report is

   A. V, IV, I, III, II
   B. IV, V, I, III, II
   C. V, I, III, II, IV
   D. IV, I, III, II, V

6. Police Officer Caldwell is completing a Complaint Report. The report will include the following five sentences:
   I. When I yelled, *Don't move, Police,* the taller man dropped the bat and ran.
   II. I asked the girl for a description of the two men.
   III. I called for an ambulance.
   IV. A young girl approached me and stated that a man with a baseball bat was beating another man in front of 1700 Grande Street.
   V. Upon approaching the location, I observed the taller man hitting the other man with the bat.

   The MOST logical order for the above sentences to appear in the report is

   A. IV, V, I, II, III
   B. V, IV, II, III, I
   C. V, I, III, IV, II
   D. IV, II, V, I, III

7. Police Officer Moore is writing a memobook entry concerning a summons he issued. The entry will contain the following five sentences:
   I. As I was walking down the platform, I heard music coming from a radio that a man was holding on his shoulder.
   II. I asked the man for some identification.
   III. I was walking in the subway when a passenger complained about a man playing a radio loudly at the opposite end of the station.
   IV. I then gave the man a summons for playing the radio. V. As soon as the man saw me approaching, he turned the radio off.

   The MOST logical order for the above sentences to appear in the memobook entry is

   A. III, V, II, I, IV
   B. I, II, V, IV, III
   C. III, I, V, II, IV
   D. I, V, II, IV, III

8. Police Officer Kashawahara is completing an Incident Report regarding fleeing suspects he had pursued earlier. The report will include the following five sentences:
   I. I saw two males attempting to break into a store through the front window.
   II. On Myrtle Avenue, they ran into an alley between two abandoned buildings.
   III. I yelled to them, *Hey, what are you guys doing by that window?*
   IV. At that time, I lost sight of the suspects and I returned to the station house.
   V. They started to run south on Wycoff Avenue heading towards Myrtle Avenue.

   The MOST logical order for the above sentences to appear in the report is

   A. I, V, II, IV, III
   B. III, V, II, IV, I
   C. I, III, V, II, IV
   D. III, I, V, II, IV

9. Police Officer Bloom is completing an entry in his memo-book regarding a confession made by a perpetrator. The entry will include the following five sentences:
   I. I went towards the dresser and took $400 in cash and a jewelry box with rings, watches, and other items in it.
   II. There in the bedroom, lying on the bed, a woman was sleeping.
   III. It was about 1:00 A.M. when I entered the apartment through an opened rear window.
   IV. I spun around, punched her in the face with my free hand, and then jumped out the window into the street.
   V. I walked back to the window carrying the money and the jewelry box and was about to go out when all of a sudden I heard the woman scream.

   The MOST logical order for the above sentences to appear in the memobook entry is

   A. I, III, II, V, IV
   B. I, V, IV, III, II
   C. III, II, I, V, IV
   D. III, V, IV, I, II

10. Police Officer Webster is preparing an Arrest Report which will include the following five sentences:
    I. I noticed that the robber had a knife placed at the victim's neck.
    II. I told the robber to drop the knife.
    III. While on patrol, I observed a robbery which was in progress.
    IV. I grabbed the robber, placed him in handcuffs, and took him to the precinct.
    V. The robber dropped the knife and tried to flee.

    The MOST logical order for the above sentences to appear in the report is

    A. I, II, V, IV, III
    B. III, I, II, V, IV
    C. III, II, IV, I, V
    D. I, III, IV, V, II

11. Police Officer Lee is preparing a report regarding someone who apparently attempted to commit suicide with a gun. The report will include the following five sentences:
    I. At the location, the woman pointed to the open door of Apartment 7L.
    II. I called for an ambulance to respond.
    III. The male had a gun in his hand and a large head wound.
    IV. A call was received from the radio dispatcher regarding a woman who heard a gunshot at 936 45th Avenue.
    V. Upon entering Apartment 7L, I saw the body of a male on the kitchen floor.

    The MOST logical order for the above sentences to appear in the report is

    A. IV, I, V, III, II
    B. I, III, V, IV, II
    C. I, V, III, II, IV
    D. IV, V, III, II, I

11.____

12. Police Officer Modrak is completing a memobook entry which will include the following five sentences:
    I. The victim, a male in his thirties, told me that the robbery occurred a few minutes ago.
    II. My partner and I jumped out of the patrol car and arrested the suspect.
    III. We responded to an armed robbery in progress at Billings Avenue and 59th Street.
    IV. On Chester Avenue and 68th Street, the victim spotted and identified the suspect.
    V. I told the victim to get into the patrol car and that we would drive him around the area.

    The MOST logical order for the above sentences to appear in the memobook is

    A. III, I, V, IV, II
    B. I, III, V, II, IV
    C. I, IV, III, V, II
    D. III, V, I, II, IV

12.____

13. Police Officer Rodriguez is preparing a report concerning an incident in which she used her revolver. Her report will include the following five sentences:
    I. Upon seeing my revolver, the robber dropped his gun to the ground.
    II. At about 10:55 P.M., I was informed by a passerby that several people were being robbed at gunpoint on 174th Street and Walton Avenue.
    III. I was assigned to patrol on 174th Street and Ghent Avenue during the evening shift.
    IV. I saw a man holding a gun on three people, took out my revolver, and shouted, *Police, don't move!*
    V. After calling for assistance, I went to 174th Street and Walton Avenue and took cover behind a car.

    The MOST logical order for the above sentences to appear in the report is

    A. II, III, IV, V, I
    B. IV, V, I, III, II
    C. III, II, V, IV, I
    D. II, IV, I, V, III

13.____

14. Police Officer Davis is completing an Activity Log entry which will include the following five sentences:
    I. A radio car was dispatched and the male was taken to Greenville Hospital.
    II. Several people saw him and called the police.
    III. A naked man was running down the street waving his arms above his head and screaming, *Insects are all over me!*
    IV. I arrived on the scene and requested an ambulance.
    V. The dispatcher informed me that no ambulances were available.

    The MOST logical order for the above sentences to appear in the Activity Log is

    A. III, IV, V, I, II
    B. II, III, V, I, IV
    C. III, II, IV, V, I
    D. II, IV, III, V, I

15. Police Officer Peake is completing an entry in his Activity Log. The entry contains the following five sentences:
    I. He went to his parked car only to find he was blocked in.
    II. The owner of the vehicle refused to move the van until he had finished his lunch.
    III. Approximately 30 minutes later, I arrived on the scene and ordered the owner of the van to remove the vehicle.
    IV. Mr. O'Neil had an appointment and was in a hurry to keep it.
    V. Mr. O'Neil entered a nearby delicatessen and asked if anyone in there drove a dark blue van, license plate number BUS 265.

    The MOST logical order for the above sentences to appear in the Activity Log is

    A. II, III, I, IV, V
    B. IV, I, V, II, III
    C. V, IV, I, III, II
    D. II, I, III, IV, V

16. Police Officer Harrison is preparing a report regarding a 10-year-old who was sexually abused at school. The report will include the following five sentences:
    I. The child described the perpetrator as a white male with a mustache, six feet tall, wearing a green uniform.
    II. On September 10, I responded to General Hospital to interview a child who was sexually abused.
    III. He later confessed at the station house.
    IV. After I interviewed the child, I responded to the school and found a janitor who fit the description.
    V. I interviewed the janitor and took him to the station house for further investigation.

    The MOST logical order for the above sentences to appear in the report is

    A. II, IV, I, V, III
    B. I, IV, V, II, III
    C. II, I, IV, V, III
    D. V, III, II, I, IV

17. Police Officer Madden is completing a report of a theft. The report will include the following five sentences:
    I. I followed behind the suspect for two blocks.
    II. I saw a man pass by the radio car carrying a shopping bag.
    III. I looked back in the direction he had just come from and noticed that the top of a parking meter was missing.
    IV. As he saw me, he started to walk faster, and I noticed a red piece of metal with the word *violation* drop out of the shopping bag.
    V. When I saw a parking meter in the shopping bag, I apprehended the suspect and placed him under arrest.

    The MOST logical order for the above sentences to appear in the report is

    A. I, IV, II, III, V
    B. II, I, IV, V, III
    C. II, IV, III, I, V
    D. III, II, IV, I, V

18. Police Officer McCaslin is preparing a report of disorderly conduct which will include the following five sentences:
    I. Police Officer Kenny and I were on patrol in a radio car when we received a dispatch to go to the Hard Rock Disco on Third Avenue.
    II. We arrived at the scene and found three men arguing loudly and obviously intoxicated.
    III. The dispatcher had received a call from a bartender regarding a dispute.
    IV. Two of the men left the disco shortly before we did.
    V. We calmed the men down after managing to separate them.

    The MOST logical order for the above sentences to appear in the report is

    A. I, II, V, III, IV
    B. III, I, IV, II, V
    C. II, I, III, IV, V
    D. I, III, II, V, IV

19. Police Officer Langhorne is completing a report of a murder. The report will contain the following five statements made by a witness:
    I. The noise created by the roar of a motorcycle caused me to look out of my window.
    II. I ran out of the house and realized the man was dead, which is when I called the police.
    III. I saw a man driving at high speed down the dead-end street on a motorcycle, closely followed by a green BMW.
    IV. The motorcyclist then parked the bike and approached the car, which was occupied by two males.
    V. Two shots were fired and the cyclist fell to the ground; then the car made a u-turn and sped down the street.

    The MOST logical order for the above sentences to appear in the report is

    A. I, II, IV, III, V
    B. V, II, I, IV, III
    C. I, III, IV, V, II
    D. III, IV, I, II, V

7 (#2)

20. Police Officer Murphy is preparing a report of a person who was assaulted. The report will include the following five sentences:
    I. I responded to the scene, but Mr. Jones had already fled.
    II. She was bleeding profusely from a cut above her right eye.
    III. Mr. and Mrs. Jones apparently were fighting in the street when Mr. Jones punched his wife in the face.
    IV. I then applied pressure to the cut to control the bleeding.
    V. I called the dispatcher on the radio to send an ambulance to respond to the scene.

    The MOST logical order for the above sentences to appear in the report is

    A. III, II, IV, I, V
    B. III, I, II, IV, V
    C. I, V, II, III, IV
    D. II, V, IV, III, I

20.____

KEY (CORRECT ANSWERS)

1. C
2. B
3. A
4. A
5. B

6. D
7. C
8. C
9. C
10. B

11. A
12. A
13. C
14. C
15. B

16. C
17. C
18. D
19. C
20. B

PHILOSOPHY, PRINCIPLES, PRACTICES AND TECHNICS OF SUPERVISION, ADMINISTRATION, MANAGEMENT AND ORGANIZATION

TABLE OF CONTENTS

		Page
I.	MEANING OF SUPERVISION	1
II.	THE OLD AND THE NEW SUPERVISION	1
III.	THE EIGHT (8) BASIC PRINCIPLES OF THE NEW SUPERVISION	1
	1. Principle of Responsibility	1
	2. Principle of Authority	2
	3. Principle of Self-Growth	2
	4. Principle of Individual Worth	2
	5. Principle of Creative Leadership	2
	6. Principle of Success and Failure	2
	7. Principle of Science	3
	8. Principle of Cooperation	3
IV.	WHAT IS ADMINISTRATION?	3
	1. Practices commonly classed as "Supervisory"	3
	2. Practices commonly classed as "Administrative"	3
	3. Practices classified as both "Supervisory" and "Administrative"	4
V.	RESPONSIBILITIES OF THE SUPERVISOR	4
VI.	COMPETENCIES OF THE SUPERVISOR	4
VII.	THE PROFESSIONAL SUPERVISOR—EMPLOYEE RELATIONSHIP	4
VIII.	MINI-TEXT IN SUPERVISION, ADMINISTRATION, MANAGEMENT AND ORGANIZATION	5
	A. Brief Highlights	5
	1. Levels of Management	5
	2. What the Supervisor Must Learn	6
	3. A Definition of Supervision	6
	4. Elements of the Team Concept	6
	5. Principles of Organization	6
	6. The Four Important Parts of Every Job	6
	7. Principles of Delegation	6
	8. Principles of Effective Communications	7
	9. Principles of Work Improvement	7

TABLE OF CONTENTS (CONTINUED)

10. Areas of Job Improvement	7
11. Seven Key Points in Making Improvements	7
12. Corrective Techniques for Job Improvement	7
13. A Planning Checklist	8
14. Five Characteristics of Good Directions	8
15. Types of Directions	8
16. Controls	8
17. Orienting the New Employee	8
18. Checklist for Orienting New Employees	8
19. Principles of Learning	9
20. Causes of Poor Performance	9
21. Four Major Steps in On-The-Job Instructions	9
22. Employees Want Five Things	9
23. Some Don'ts in Regard to Praise	9
24. How to Gain Your Workers' Confidence	9
25. Sources of Employee Problems	9
26. The Supervisor's Key to Discipline	10
27. Five Important Processes of Management	10
28. When the Supervisor Fails to Plan	10
29. Fourteen General Principles of Management	10
30. Change	10
B. Brief Topical Summaries	11
I. Who/What is the Supervisor?	11
II. The Sociology of Work	11
III. Principles and Practices of Supervision	12
IV. Dynamic Leadership	12
V. Processes for Solving Problems	12
VI. Training for Results	13
VII. Health, Safety and Accident Prevention	13
VIII. Equal Employment Opportunity	13
IX. Improving Communications	14
X. Self-Development	14
XI. Teaching and Training	14
A. The Teaching Process	14
1. Preparation	14
2. Presentation	15
3. Summary	15
4. Application	15
5. Evaluation	15
B. Teaching Methods	15
1. Lecture	15
2. Discussion	15
3. Demonstration	16
4. Performance	16
5. Which Method to Use	16

PHILOSOPHY, PRINCIPLES, PRACTICES, AND TECHNICS
OF
SUPERVISION, ADMINISTRATION, MANAGEMENT AND ORGANIZATION

I. MEANING OF SUPERVISION

The extension of the democratic philosophy has been accompanied by an extension in the scope of supervision. Modern leaders and supervisors no longer think of supervision in the narrow sense of being confined chiefly to visiting employees, supplying materials, or rating the staff. They regard supervision as being intimately related to all the concerned agencies of society, they speak of the supervisor's function in terms of "growth", rather than the "improvement," of employees.

This modern concept of supervision may be defined as follows:

Supervision is leadership and the development of leadership within groups which are cooperatively engaged in inspection, research, training, guidance and evaluation.

II. THE OLD AND THE NEW SUPERVISION

TRADITIONAL
1. Inspection
2. Focused on the employee
3. Visitation
4. Random and haphazard
5. Imposed and authoritarian
6. One person usually

MODERN
1. Study and analysis
2. Focused on aims, materials, methods, supervisors, employees, environment
3. Demonstrations, intervisitation, workshops, directed reading, bulletins, etc.
4. Definitely organized and planned (scientific)
5. Cooperative and democratic
6. Many persons involved (creative)

III THE EIGHT (8) BASIC PRINCIPLES OF THE NEW SUPERVISION

1. *PRINCIPLE OF RESPONSIBILITY*
Authority to act and responsibility for acting must be joined.
   a. If you give responsibility, give authority.
   b. Define employee duties clearly.
   c. Protect employees from criticism by others.
   d. Recognize the rights as well as obligations of employees.
   e. Achieve the aims of a democratic society insofar as it is possible within the area of your work.
   f. Establish a situation favorable to training and learning.
   g. Accept ultimate responsibility for everything done in your section, unit, office, division, department.
   h. Good administration and good supervision are inseparable.

2. PRINCIPLE OF AUTHORITY

The success of the supervisor is measured by the extent to which the power of authority is not used.

   a. Exercise simplicity and informality in supervision.
   b. Use the simplest machinery of supervision.
   c. If it is good for the organization as a whole, it is probably justified.
   d. Seldom be arbitrary or authoritative.
   e. Do not base your work on the power of position or of personality.
   f. Permit and encourage the free expression of opinions.

3. PRINCIPLE OF SELF-GROWTH

The success of the supervisor is measured by the extent to which, and the speed with which, he is no longer needed.

   a. Base criticism on principles, not on specifics.
   b. Point out higher activities to employees.
   c. Train for self-thinking by employees, to meet new situations.
   d. Stimulate initiative, self-reliance and individual responsibility.
   e. Concentrate on stimulating the growth of employees rather than on removing defects.

4. PRINCIPLE OF INDIVIDUAL WORTH

Respect for the individual is a paramount consideration in supervision.

   a. Be human and sympathetic in dealing with employees.
   b. Don't nag about things to be done.
   c. Recognize the individual differences among employees and seek opportunities to permit best expression of each personality.

5. PRINCIPLE OF CREATIVE LEADERSHIP

The best supervision is that which is not apparent to the employee.

   a. Stimulate, don't drive employees to creative action.
   b. Emphasize doing good things.
   c. Encourage employees to do what they do best.
   d. Do not be too greatly concerned with details of subject or method.
   e. Do not be concerned exclusively with immediate problems and activities.
   f. Reveal higher activities and make them both desired and maximally possible.
   g. Determine procedures in the light of each situation but see that these are derived from a sound basic philosophy.
   h. Aid, inspire and lead so as to liberate the creative spirit latent in all good employees.

6. PRINCIPLE OF SUCCESS AND FAILURE

There are no unsuccessful employees, only unsuccessful supervisors who have failed to give proper leadership.

   a. Adapt suggestions to the capacities, attitudes, and prejudices of employees.
   b. Be gradual, be progressive, be persistent.
   c. Help the employee find the general principle; have the employee apply his own problem to the general principle.
   d. Give adequate appreciation for good work and honest effort.
   e. Anticipate employee difficulties and help to prevent them.
   f. Encourage employees to do the desirable things they will do anyway.
   g. Judge your supervision by the results it secures.

7. PRINCIPLE OF SCIENCE

Successful supervision is scientific, objective, and experimental. It is based on facts, not on prejudices.
      a. Be cumulative in results.
      b. Never divorce your suggestions from the goals of training.
      c. Don't be impatient of results.
      d. Keep all matters on a professional, not a personal level.
      e. Do not be concerned exclusively with immediate problems and activities.
      f. Use objective means of determining achievement and rating where possible.

8. PRINCIPLE OF COOPERATION

Supervision is a cooperative enterprise between supervisor and employee.
      a. Begin with conditions as they are.
      b. Ask opinions of all involved when formulating policies.
      c. Organization is as good as its weakest link.
      d. Let employees help to determine policies and department programs.
      e. Be approachable and accessible - physically and mentally.
      f. Develop pleasant social relationships.

IV. WHAT IS ADMINISTRATION?

Administration is concerned with providing the environment, the material facilities, and the operational procedures that will promote the maximum growth and development of supervisors and employees. (Organization is an aspect, and a concomitant, of administration.)

There is no sharp line of demarcation between supervision and administration; these functions are intimately interrelated and, often, overlapping. They are complementary activities.

1. PRACTICES COMMONLY CLASSED AS "SUPERVISORY"
      a. Conducting employees conferences
      b. Visiting sections, units, offices, divisions, departments
      c. Arranging for demonstrations
      d. Examining plans
      e. Suggesting professional reading
      f. Interpreting bulletins
      g. Recommending in-service training courses
      h. Encouraging experimentation
      i. Appraising employee morale
      j. Providing for intervisitation

2. PRACTICES COMMONLY CLASSIFIED AS "ADMINISTRATIVE"
      a. Management of the office
      b. Arrangement of schedules for extra duties
      c. Assignment of rooms or areas
      d. Distribution of supplies
      e. Keeping records and reports
      f. Care of audio-visual materials
      g. Keeping inventory records
      h. Checking record cards and books
      i. Programming special activities
      j. Checking on the attendance and punctuality of employees

3. *PRACTICES COMMONLY CLASSIFIED AS BOTH "SUPERVISORY" AND "ADMINISTRATIVE"*
    a. Program construction
    b. Testing or evaluating outcomes
    c. Personnel accounting
    d. Ordering instructional materials

V. RESPONSIBILITIES OF THE SUPERVISOR

A person employed in a supervisory capacity must constantly be able to improve his own efficiency and ability. He represents the employer to the employees and only continuous self-examination can make him a capable supervisor.

Leadership and training are the supervisor's responsibility. An efficient working unit is one in which the employees work with the supervisor. It is his job to bring out the best in his employees. He must always be relaxed, courteous and calm in his association with his employees. Their feelings are important, and a harsh attitude does not develop the most efficient employees.

VI. COMPETENCIES OF THE SUPERVISOR

1. Complete knowledge of the duties and responsibilities of his position.
2. To be able to organize a job, plan ahead and carry through.
3. To have self-confidence and initiative.
4. To be able to handle the unexpected situation and make quick decisions.
5. To be able to properly train subordinates in the positions they are best suited for.
6. To be able to keep good human relations among his subordinates.
7. To be able to keep good human relations between his subordinates and himself and to earn their respect and trust.

VII. THE PROFESSIONAL SUPERVISOR-EMPLOYEE RELATIONSHIP

There are two kinds of efficiency: one kind is only apparent and is produced in organizations through the exercise of mere discipline; this is but a simulation of the second, or true, efficiency which springs from spontaneous cooperation. If you are a manager, no matter how great or small your responsibility, it is your job, in the final analysis, to create and develop this involuntary cooperation among the people whom you supervise. For, no matter how powerful a combination of money, machines, and materials a company may have, this is a dead and sterile thing without a team of willing, thinking and articulate people to guide it.

The following 21 points are presented as indicative of the exemplary basic relationship that should exist between supervisor and employee:

1. Each person wants to be liked and respected by his fellow employee and wants to be treated with consideration and respect by his superior.
2. The most competent employee will make an error. However, in a unit where good relations exist between the supervisor and his employees, tenseness and fear do not exist. Thus, errors are not hidden or covered up and the efficiency of a unit is not impaired.
3. Subordinates resent rules, regulations, or orders that are unreasonable or unexplained.
4. Subordinates are quick to resent unfairness, harshness, injustices and favoritism.
5. An employee will accept responsibility if he knows that he will be complimented for a job well done, and not too harshly chastised for failure; that his supervisor will check the cause of the failure, and, if it was the supervisor's fault, he will assume the blame therefore. If it was the employee's fault, his supervisor will explain the correct method or means of handling the responsibility.

6. An employee wants to receive credit for a suggestion he has made, that is used. If a suggestion cannot be used, the employee is entitled to an explanation. The supervisor should not say "no" and close the subject.
7. Fear and worry slow up a worker's ability. Poor working environment can impair his physical and mental health. A good supervisor avoids forceful methods, threats and arguments to get a job done.
8. A forceful supervisor is able to train his employees individually and as a team, and is able to motivate them in the proper channels.
9. A mature supervisor is able to properly evaluate his subordinates and to keep them happy and satisfied.
10. A sensitive supervisor will never patronize his subordinates.
11. A worthy supervisor will respect his employees' confidences.
12. Definite and clear-cut responsibilities should be assigned to each executive.
13. Responsibility should always be coupled with corresponding authority.
14. No change should be made in the scope or responsibilities of a position without a definite understanding to that effect on the part of all persons concerned.
15. No executive or employee, occupying a single position in the organization, should be subject to definite orders from more than one source.
16. Orders should never be given to subordinates over the head of a responsible executive. Rather than do this, the officer in question should be supplanted.
17. Criticisms of subordinates should, whoever possible, be made privately, and in no case should a subordinate be criticized in the presence of executives or employees of equal or lower rank.
18. No dispute or difference between executives or employees as to authority or responsibilities should be considered too trivial for prompt and careful adjudication.
19. Promotions, wage changes, and disciplinary action should always be approved by the executive immediately superior to the one directly responsible.
20. No executive or employee should ever be required, or expected, to be at the same time an assistant to, and critic of, another.
21. Any executive whose work is subject to regular inspection should, whever practicable, be given the assistance and facilities necessary to enable him to maintain an independent check of the quality of his work.

VIII. MINI-TEXT IN SUPERVISION, ADMINISTRATION, MANAGEMENT, AND ORGANIZATION

A. BRIEF HIGHLIGHTS

Listed concisely and sequentially are major headings and important data in the field for quick recall and review.

1. *LEVELS OF MANAGEMENT*

Any organization of some size has several levels of management. In terms of a ladder the levels are:

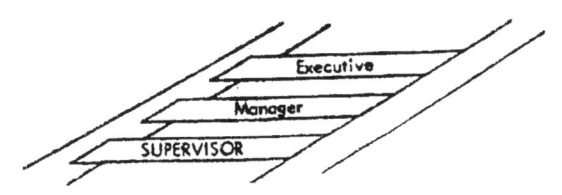

```
          Executive
        Manager
     SUPERVISOR
```

The first level is very important because it is the beginning point of management leadership.

2. WHAT THE SUPERVISOR MUST LEARN
A supervisor must learn to:
(1) Deal with people and their differences
(2) Get the job done through people
(3) Recognize the problems when they exist
(4) Overcome obstacles to good performance
(5) Evaluate the performance of people
(6) Check his own performance in terms of accomplishment

3. A DEFINITION OF SUPERVISOR
The term supervisor means any individual having authority, in the interests of the employer, to hire, transfer, suspend, lay-off, recall, promote, discharge, assign, reward, or discipline other employees or responsibility to direct them, or to adjust their grievances, or effectively to recommend such action, if, in connection with the foregoing, exercise of such authority is not of a merely routine or clerical nature but requires the use of independent judgment.

4. ELEMENTS OF THE TEAM CONCEPT
What is involved in teamwork? The component parts are:

(1) Members	(3) Goals	(5) Cooperation
(2) A leader	(4) Plans	(6) Spirit

5. PRINCIPLES OF ORGANIZATION
(1) A team member must know what his job is.
(2) Be sure that the nature and scope of a job are understood.
(3) Authority and responsibility should be carefully spelled out.
(4) A supervisor should be permitted to make the maximum number of decisions affecting his employees.
(5) Employees should report to only one supervisor.
(6) A supervisor should direct only as many employees as he can handle effectively.
(7) An organization plan should be flexible.
(8) Inspection and performance of work should be separate.
(9) Organizational problems should receive immediate attention.
(10) Assign work in line with ability and experience.

6. THE FOUR IMPORTANT PARTS OF EVERY JOB
(1) Inherent in every job is the *accountability* for results.
(2) A second set of factors in every job is *responsibilities.*
(3) Along with duties and responsibilities one must have the *authority* to act within certain limits without obtaining permission to proceed.
(4) No job exists in a vacuum. The supervisor is surrounded by key *relationships.*

7. PRINCIPLES OF DELEGATION
Where work is delegated for the first time, the supervisor should think in terms of these questions:
(1) Who is best qualified to do this?
(2) Can an employee improve his abilities by doing this?
(3) How long should an employee spend on this?
(4) Are there any special problems for which he will need guidance?
(5) How broad a delegation can I make?

8. PRINCIPLES OF EFFECTIVE COMMUNICATIONS
   (1) Determine the media
   (2) To whom directed?
   (3) Identification and source authority
   (4) Is communication understood?

9. PRINCIPLES OF WORK IMPROVEMENT
   (1) Most people usually do only the work which is assigned to them
   (2) Workers are likely to fit assigned work into the time available to perform it
   (3) A good workload usually stimulates output
   (4) People usually do their best work when they know that results will be reviewed or inspected
   (5) Employees usually feel that someone else is responsible for conditions of work, workplace layout, job methods, type of tools/equipment, and other such factors
   (6) Employees are usually defensive about their job security
   (7) Employees have natural resistance to change
   (8) Employees can support or destroy a supervisor
   (9) A supervisor usually earns the respect of his people through his personal example of diligence and efficiency

10. AREAS OF JOB IMPROVEMENT
The areas of job improvement are quite numerous, but the most common ones which a supervisor can identify and utilize are:
   (1) Departmental layout   (5) Work methods
   (2) Flow of work   (6) Materials handling
   (3) Workplace layout   (7) Utilization
   (4) Utilization of manpower (8) Motion economy

11. SEVEN KEY POINTS IN MAKING IMPROVEMENTS
   (1) Select the job to be improved
   (2) Study how it is being done now
   (3) Question the present method
   (4) Determine actions to be taken
   (5) Chart proposed method
   (6) Get approval and apply
   (7) Solicit worker participation

12. CORRECTIVE TECHNIQUES OF JOB IMPROVEMENT

Specific Problems	General Improvement	Corrective Techniques
(1) Size of workload	(1) Departmental layout	(1) Study with scale model
(2) Inability to meet schedules	(2) Flow of work	(2) Flow chart study
(3) Strain and fatigue	(3) Work plan layout	(3) Motion analysis
(4) Improper use of men and skills	(4) Utilization of manpower	(4) Comparison of units produced to standard allowance
(5) Waste, poor quality, unsafe conditions	(5) Work methods	(5) Methods analysis
(6) Bottleneck conditions that hinder output	(6) Materials handling	(6) Flow chart & equipment study
(7) Poor utilization of equipment and machine	(7) Utilization of equipment	(7) Down time vs. running time
(8) Efficiency and productivity of labor	(8) Motion economy	(8) Motion analysis

13. A *PLANNING CHECKLIST*
    - (1) Objectives
    - (2) Controls
    - (3) Delegations
    - (4) Communications
    - (5) Resources
    - (6) Resources
    - (7) Manpower
    - (8) Equipment
    - (9) Supplies and materials
    - (10) Utilization of time
    - (11) Safety
    - (12) Money
    - (13) Work
    - (14) Timing of improvements

14. *FIVE CHARACTERISTICS OF GOOD DIRECTIONS*
    In order to get results, directions must be:
    - (1) Possible of accomplishment
    - (2) Agreeable with worker interests
    - (3) Related to mission
    - (4) Planned and complete
    - (5) Unmistakably clear

15. *TYPES OF DIRECTIONS*
    - (1) Demands or direct orders
    - (2) Requests
    - (3) Suggestion or implication
    - (4) Volunteering

16. *CONTROLS*
    A typical listing of the overall areas in which the supervisor should establish controls might be:
    - (1) Manpower
    - (2) Materials
    - (3) Quality of work
    - (4) Quantity of work
    - (5) Time
    - (6) Space
    - (7) Money
    - (8) Methods

17. *ORIENTING THE NEW EMPLOYEE*
    - (1) Prepare for him
    - (2) Welcome the new employee
    - (3) Orientation for the job
    - (4) Follow-up

18. *CHECKLIST FOR ORIENTING NEW EMPLOYEES*   Yes   No
    - (1) Do your appreciate the feelings of new employees when they first report for work?   ____ ____
    - (2) Are you aware of the fact that the new employee must make a big adjustment to his job?   ____ ____
    - (3) Have you given him good reasons for liking the job and the organization?   ____ ____
    - (4) Have you prepared for his first day on the job?
    - (5) Did you welcome him cordially and make him feel needed?
    - (6) Did you establish rapport with him so that he feels free to talk and discuss matters with you?
    - (7) Did you explain his job to him and his relationship to you?   ____ ____
    - (8) Does he know that his work will be evaluated periodically on a basis that is fair and objective?   ____ ____
    - (9) Did you introduce him to his fellow workers in such a way that they are likely to accept him?   ____ ____
    - (10) Does he know what employee benefits he will receive?
    - (11) Does he understand the importance of being on the job and what to do if he must leave his duty station?   ____ ____
    - (12) Has he been impressed with the importance of accident prevention and safe practice?   ____ ____
    - (13) Does he generally know his way around the department?   ____ ____
    - (14) Is he under the guidance of a sponsor who will teach the right ways of doing things?   ____ ____
    - (15) Do you plan to follow-up so that he will continue to adjust successfully to his job?   ____ ____

19. *PRINCIPLES OF LEARNING*
    (1) Motivation   (2) Demonstration or explanation   (3) Practice

20. *CAUSES OF POOR PERFORMANCE*
    (1) Improper training for job
    (2) Wrong tools
    (3) Inadequate directions
    (4) Lack of supervisory follow-up
    (5) Poor communications
    (6) Lack of standards of performance
    (7) Wrong work habits
    (8) Low morale
    (9) Other

21. *FOUR MAJOR STEPS IN ON-THE-JOB INSTRUCTION*
    (1) Prepare the worker
    (2) Present the operation
    (3) Tryout performance
    (4) Follow-up

22. *EMPLOYEES WANT FIVE THINGS*
    (1) Security (2) Opportunity (3) Recognition (4) Inclusion (5) Expression

23. *SOME DON'TS IN REGARD TO PRAISE*
    (1) Don't praise a person for something he hasn't done
    (2) Don't praise a person unless you can be sincere
    (3) Don't be sparing in praise just because your superior withholds it from you
    (4) Don't let too much time elapse between good performance and recognition of it

24. *HOW TO GAIN YOUR WORKERS' CONFIDENCE*
    Methods of developing confidence include such things as:
    (1) Knowing the interests, habits, hobbies of employees
    (2) Admitting your own inadequacies
    (3) Sharing and telling of confidence in others
    (4) Supporting people when they are in trouble
    (5) Delegating matters that can be well handled
    (6) Being frank and straightforward about problems and working conditions
    (7) Encouraging others to bring their problems to you
    (8) Taking action on problems which impede worker progress

25. *SOURCES OF EMPLOYEE PROBLEMS*
    On-the-job causes might be such things as:
    (1) A feeling that favoritism is exercised in assignments
    (2) Assignment of overtime
    (3) An undue amount of supervision
    (4) Changing methods or systems
    (5) Stealing of ideas or trade secrets
    (6) Lack of interest in job
    (7) Threat of reduction in force
    (8) Ignorance or lack of communications
    (9) Poor equipment
    (10) Lack of knowing how supervisor feels toward employee
    (11) Shift assignments

    Off-the-job problems might have to do with:
    (1) Health   (2) Finances   (3) Housing   (4) Family

26. THE SUPERVISOR'S KEY TO DISCIPLINE

There are several key points about discipline which the supervisor should keep in mind:
(1) Job discipline is one of the disciplines of life and is directed by the supervisor.
(2) It is more important to correct an employee fault than to fix blame for it.
(3) Employee performance is affected by problems both on the job and off.
(4) Sudden or abrupt changes in behavior can be indications of important employee problems.
(5) Problems should be dealt with as soon as possible after they are identified.
(6) The attitude of the supervisor may have more to do with solving problems than the techniques of problem solving.
(7) Correction of employee behavior should be resorted to only after the supervisor is sure that training or counseling will not be helpful.
(8) Be sure to document your disciplinary actions.
(9) Make sure that you are disciplining on the basis of facts rather than personal feelings.
(10) Take each disciplinary step in order, being careful not to make snap judgments, or decisions based on impatience.

27. FIVE IMPORTANT PROCESSES OF MANAGEMENT

(1) Planning   (2) Organizing   (3) Scheduling
(4) Controlling   (5) Motivating

28. WHEN THE SUPERVISOR FAILS TO PLAN

(1) Supervisor creates impression of not knowing his job
(2) May lead to excessive overtime
(3) Job runs itself -- supervisor lacks control
(4) Deadlines and appointments missed
(5) Parts of the work go undone
(6) Work interrupted by emergencies
(7) Sets a bad example
(8) Uneven workload creates peaks and valleys
(9) Too much time on minor details at expense of more important tasks

29. FOURTEEN GENERAL PRINCIPLES OF MANAGEMENT

(1) Division of work
(2) Authority and responsibility
(3) Discipline
(4) Unity of command
(5) Unity of direction
(6) Subordination of individual interest to general interest
(7) Remuneration of personnel
(8) Centralization
(9) Scalar chain
(10) Order
(11) Equity
(12) Stability of tenure of personnel
(13) Initiative
(14) Esprit de corps

30. CHANGE

Bringing about change is perhaps attempted more often, and yet less well understood, than anything else the supervisor does. How do people generally react to change? (People tend to resist change that is imposed upon them by other individuals or circumstances.

Change is characteristic of every situation. It is a part of every real endeavor where the efforts of people are concerned.

A. Why do people resist change?
   People may resist change because of:
   (1) Fear of the unknown
   (2) Implied criticism
   (3) Unpleasant experiences in the past
   (4) Fear of loss of status
   (5) Threat to the ego
   (6) Fear of loss of economic stability

B. How can we best overcome the resistance to change?
   In initiating change, take these steps:
   (1) Get ready to sell
   (2) Identify sources of help
   (3) Anticipate objections
   (4) Sell benefits
   (5) Listen in depth
   (6) Follow up

B. BRIEF TOPICAL SUMMARIES

I. WHO/WHAT IS THE SUPERVISOR?
1. The supervisor is often called the "highest level employee and the lowest level manager."
2. A supervisor is a member of both management and the work group. He acts as a bridge between the two.
3. Most problems in supervision are in the area of human relations, or people problems.
4. Employees expect: Respect, opportunity to learn and to advance, and a sense of belonging, and so forth.
5. Supervisors are responsible for directing people and organizing work. Planning is of paramount importance.
6. A position description is a set of duties and responsibilities inherent to a given position.
7. It is important to keep the position description up-to-date and to provide each employee with his own copy.

II. THE SOCIOLOGY OF WORK
1. People are alike in many ways; however, each individual is unique.
2. The supervisor is challenged in getting to know employee differences. Acquiring skills in evaluating individuals is an asset.
3. Maintaining meaningful working relationships in the organization is of great importance.
4. The supervisor has an obligation to help individuals to develop to their fullest potential.
5. Job rotation on a planned basis helps to build versatility and to maintain interest and enthusiasm in work groups.
6. Cross training (job rotation) provides backup skills.
7. The supervisor can help reduce tension by maintaining a sense of humor, providing guidance to employees, and by making reasonable and timely decisions. Employees respond favorably to working under reasonably predictable circumstances.
8. Change is characteristic of all managerial behavior. The supervisor must adjust to changes in procedures, new methods, technological changes, and to a number of new and sometimes challenging situations.
9. To overcome the natural tendency for people to resist change, the supervisor should become more skillful in initiating change.

III. PRINCIPLES AND PRACTICES OF SUPERVISION
1. Employees should be required to answer to only one superior.
2. A supervisor can effectively direct only a limited number of employees, depending upon the complexity, variety, and proximity of the jobs involved.
3. The organizational chart presents the organization in graphic form. It reflects lines of authority and responsibility as well as interrelationships of units within the organization.
4. Distribution of work can be improved through an analysis using the "Work Distribution Chart."
5. The "Work Distribution Chart" reflects the division of work within a unit in understandable form.
6. When related tasks are given to an employee, he has a better chance of increasing his skills through training.
7. The individual who is given the responsibility for tasks must also be given the appropriate authority to insure adequate results.
8. The supervisor should delegate repetitive, routine work. Preparation of recurring reports, maintaining leave and attendance records are some examples.
9. Good discipline is essential to good task performance. Discipline is reflected in the actions of employees on the job in the absence of supervision.
10. Disciplinary action may have to be taken when the positive aspects of discipline have failed. Reprimand, warning, and suspension are examples of disciplinary action.
11. If a situation calls for a reprimand, be sure it is deserved and remember it is to be done in private.

IV. DYNAMIC LEADERSHIP
1. A style is a personal method or manner of exerting influence.
2. Authoritarian leaders often see themselves as the source of power and authority.
3. The democratic leader often perceives the group as the source of authority and power.
4. Supervisors tend to do better when using the pattern of leadership that is most natural for them.
5. Social scientists suggest that the effective supervisor use the leadership style that best fits the problem or circumstances involved.
6. All four styles -- telling, selling, consulting, joining -- have their place. Using one does not preclude using the other at another time.
7. The theory X point of view assumes that the average person dislikes work, will avoid it whenever possible, and must be coerced to achieve organizational objectives.
8. The theory Y point of view assumes that the average person considers work to be as natural as play, and, when the individual is committed, he requires little supervision or direction to accomplish desired objectives.
9. The leader's basic assumptions concerning human behavior and human nature affect his actions, decisions, and other managerial practices.
10. Dissatisfaction among employees is often present, but difficult to isolate. The supervisor should seek to weaken dissatisfaction by keeping promises, being sincere and considerate, keeping employees informed, and so forth.
11. Constructive suggestions should be encouraged during the natural progress of the work.

V. PROCESSES FOR SOLVING PROBLEMS
1. People find their daily tasks more meaningful and satisfying when they can improve them.
2. The causes of problems, or the key factors, are often hidden in the background. Ability to solve problems often involves the ability to isolate them from their backgrounds. There is some substance to the cliché that some persons "can't see the forest for the trees."
3. New procedures are often developed from old ones. Problems should be broken down into manageable parts. New ideas can be adapted from old ones.

4. People think differently in problem-solving situations. Using a logical, patterned approach is often useful. One approach found to be useful includes these steps:
- (a) Define the problem
- (b) Establish objectives
- (c) Get the facts
- (d) Weigh and decide
- (e) Take action
- (f) Evaluate action

VI. TRAINING FOR RESULTS
1. Participants respond best when they feel training is important to them.
2. The supervisor has responsibility for the training and development of those who report to him.
3. When training is delegated to others, great care must be exercised to insure the trainer has knowledge, aptitude, and interest for his work as a trainer.
4. Training (learning) of some type goes on continually. The most successful supervisor makes certain the learning contributes in a productive manner to operational goals.
5. New employees are particularly susceptible to training. Older employees facing new job situations require specific training, as well as having need for development and growth opportunities.
6. Training needs require continuous monitoring.
7. The training officer of an agency is a professional with a responsibility to assist supervisors in solving training problems.
8. Many of the self-development steps important to the supervisor's own growth are equally important to the development of peers and subordinates. Knowledge of these is important when the supervisor consults with others on development and growth opportunities.

VII. HEALTH, SAFETY, AND ACCIDENT PREVENTION
1. Management-minded supervisors take appropriate measures to assist employees in maintaining health and in assuring safe practices in the work environment.
2. Effective safety training and practices help to avoid injury and accidents.
3. Safety should be a management goal. All infractions of safety which are observed should be corrected without exception.
4. Employees' safety attitude, training and instruction, provision of safe tools and equipment, supervision, and leadership are considered highly important factors which contribute to safety and which can be influenced directly by supervisors.
5. When accidents do occur they should be investigated promptly for very important reasons, including the fact that information which is gained can be used to prevent accidents in the future.

VIII. EQUAL EMPLOYMENT OPPORTUNITY
1. The supervisor should endeavor to treat all employees fairly, without regard to religion, race, sex, or national origin.
2. Groups tend to reflect the attitude of the leader. Prejudice can be detected even in very subtle form. Supervisors must strive to create a feeling of mutual respect and confidence in every employee.
3. Complete utilization of all human resources is a national goal. Equitable consideration should be accorded women in the work force, minority-group members, the physically and mentally handicapped, and the older employee. The important question is: "Who can do the job?"
4. Training opportunities, recognition for performance, overtime assignments, promotional opportunities, and all other personnel actions are to be handled on an equitable basis.

IX. IMPROVING COMMUNICATIONS

1. Communications is achieving understanding between the sender and the receiver of a message. It also means sharing information -- the creation of understanding.
2. Communication is basic to all human activity. Words are means of conveying meanings; however, real meanings are in people.
3. There are very practical differences in the effectiveness of one-way, impersonal, and two-way communications. Words spoken face-to-face are better understood. Telephone conversations are effective, but lack the rapport of person-to-person exchanges. The whole person communicates.
4. Cooperation and communication in an organization go hand in hand. When there is a mutual respect between people, spelling out rules and procedures for communicating is unnecessary.
5. There are several barriers to effective communications. These include failure to listen with respect and understanding, lack of skill in feedback, and misinterpreting the meanings of words used by the speaker. It is also common practice to listen to what we want to hear, and tune out things we do not want to hear.
6. Communication is management's chief problem. The supervisor should accept the challenge to communicate more effectively and to improve interagency and intra-agency communications.
7. The supervisor may often plan for and conduct meetings. The planning phase is critical and may determine the success or the failure of a meeting.
8. Speaking before groups usually requires extra effort. Stage fright may never disappear completely, but it can be controlled.

X. SELF-DEVELOPMENT

1. Every employee is responsible for his own self-development.
2. Toastmaster and toastmistress clubs offer opportunities to improve skills in oral communications.
3. Planning for one's own self-development is of vital importance. Supervisors know their own strengths and limitations better than anyone else.
4. Many opportunities are open to aid the supervisor in his developmental efforts, including job assignments; training opportunities, both governmental and non-governmental -- to include universities and professional conferences and seminars.
5. Programmed instruction offers a means of studying at one's own rate.
6. Where difficulties may arise from a supervisor's being away from his work for training, he may participate in televised home study or correspondence courses to meet his self-develop- ment needs.

XI. TEACHING AND TRAINING

A. The Teaching Process

Teaching is encouraging and guiding the learning activities of students toward established goals. In most cases this process consists in five steps: preparation, presentation, summarization, evaluation, and application.

1. Preparation

   Preparation is twofold in nature; that of the supervisor and the employee.
   Preparation by the supervisor is absolutely essential to success. He must know what, when, where, how, and whom he will teach. Some of the factors that should be considered are:
   - (1) The objectives
   - (2) The materials needed
   - (3) The methods to be used
   - (4) Employee participation
   - (5) Employee interest
   - (6) Training aids
   - (7) Evaluation
   - (8) Summarization

Employee preparation consists in preparing the employee to receive the material. Probably the most important single factor in the preparation of the employee is arousing and maintaining his interest. He must know the objectives of the training, why he is there, how the material can be used, and its importance to him.

2. Presentation

In presentation, have a carefully designed plan and follow it.
The plan should be accurate and complete, yet flexible enough to meet situations as they arise. The method of presentation will be determined by the particular situation and objectives.

3. Summary

A summary should be made at the end of every training unit and program. In addition, there may be internal summaries depending on the nature of the material being taught. The important thing is that the trainee must always be able to understand how each part of the new material relates to the whole.

4. Application

The supervisor must arrange work so the employee will be given a chance to apply new knowledge or skills while the material is still clear in his mind and interest is high. The trainee does not really know whether he has learned the material until he has been given a chance to apply it. If the material is not applied, it loses most of its value.

5. Evaluation

The purpose of all training is to promote learning. To determine whether the training has been a success or failure, the supervisor must evaluate this learning.

In the broadest sense evaluation includes all the devices, methods, skills, and techniques used by the supervisor to keep him self and the employees informed as to their progress toward the objectives they are pursuing. The extent to which the employee has mastered the knowledge, skills, and abilities, or changed his attitudes, as determined by the program objectives, is the extent to which instruction has succeeded or failed.

Evaluation should not be confined to the end of the lesson, day, or program but should be used continuously. We shall note later the way this relates to the rest of the teaching process.

B. Teaching Methods

A teaching method is a pattern of identifiable student and instructor activity used in presenting training material.

All supervisors are faced with the problem of deciding which method should be used at a given time.

As with all methods, there are certain advantages and disadvantages to each method.

1. Lecture

The lecture is direct oral presentation of material by the supervisor. The present trend is to place less emphasis on the trainer's activity and more on that of the trainee.

2. Discussion

Teaching by discussion or conference involves using questions and other techniques to arouse interest and focus attention upon certain areas, and by doing so creating a learning situation. This can be one of the most valuable methods because it gives the employees 'an opportunity to express their ideas and pool their knowledge.

3. Demonstration

The demonstration is used to teach how something works or how to do something. It can be used to show a principle or what the results of a series of actions will be. A well-staged demonstration is particularly effective because it shows proper methods of performance in a realistic manner.

4. Performance

Performance is one of the most fundamental of all learning techniques or teaching methods. The trainee may be able to tell how a specific operation should be performed but he cannot be sure he knows how to perform the operation until he has done so.

5. Which Method to Use

Moreover, there are other methods and techniques of teaching. It is difficult to use any method without other methods entering into it. In any learning situation a combination of methods is usually more effective than anyone method alone.

Finally, evaluation must be integrated into the other aspects of the teaching-learning process.

It must be used in the motivation of the trainees; it must be used to assist in developing understanding during the training; and it must be related to employee application of the results of training.

This is distinctly the role of the supervisor.

BASIC FUNDAMENTALS OF OCCUPATIONAL SAFETY AND HEALTH ORGANIZATION

Analysis of safety and health programs in organizations or companies with outstanding records shows that invariably the most successful programs are built around these seven elements:

I. MANAGEMENT LEADERSHIP
    Responsibility
    Policy

II. ASSIGNMENT OF AUTHORITY
    Safety and Health Directors
    Safety and Health Committees
    Small Plant Organizations
    Scattered Operations

III. MAINTENANCE OF SAFE AND HEALTHFUL WORKING CONDITIONS
    Inspection of Work Areas
    Fire Inspections
    Health Surveys
    Job Safety Analysis

IV. ESTABLISHMENT OF SAFETY AND HEALTH TRAINING
    Employee
    Supervisor
    Job Instruction Training

V. ACCIDENT RECORD/DATA COLLECTION SYSTEM
    Records
    Accident Investigation
    Accident Analysis
    Rates
    Countermeasures

VI. HEALTH, MEDICAL AND FIRST AID SYSTEMS
    Health
    Medical
    First Aid

VII. ACCEPTANCE OF PERSONAL ACCOUNTABILITY BY EMPLOYEES
    Maintaining Interest

These seven elements of accident prevention are the same in any industry regardless of the operation and in any establishment or plant, large or small.

I. <u>MANAGEMENT LEADERSHIP</u>

<u>Responsibility</u>

Top management's attitude toward accident prevention in any company or business is almost invariably reflected in the attitude of the supervisory force. Similarly, the employees' attitude is usually the same as the supervisors'. Thus, if the top executive is not genuinely interested in preventing accidents, injuries, and occupational illnesses, no one else is likely to be. Since this basic fact applies to every level of management and supervision, an accident prevention program must have top management's personal commitment and a demonstrated interest if employee cooperation and participation are to be obtained.

Policy

To initiate the program, top management must issue a clear-cut statement of policy for the guidance of middle management, supervisors, and employees. Such a statement of policy will indicate top management's viewpoint in principle, and should cover in general the basic elements.

The details for carrying out an accident prevention program may be assigned, but the responsibility for the basic policy cannot be delegated.

Concern for the safety and health of a firm's employees doesn't stop here but also requires its active interest and participation in all of the major elements of the safety and health organization.

II. ASSIGNMENT OF AUTHORITY

Safety and Health Directors

Safety activities, like any other phase of business, must have leadership and guidance. It is of paramount importance that management assign the authority to direct the safety and health program to one individual. The individual must be formally trained, or the training must be provided on the job in the field of occupational safety and health. The title may be the safety and health director or engineer, the safety manager, or the safety supervisor, depending upon the organization, the nature of the duties assigned, and the personal qualifications. While the occupational safety and health director's exact role varies, the job usually covers:
- Developing and implementing a safety and health program.
- Identifying and controlling hazards.
- Advising management on conformance with company policy and government safety regulations.
- Helping employees understand their safety responsibilities and practices (working through the first line supervisor).
- Evaluating the severity and causes of accidents.
- Evaluating the effectiveness of the safety and health program and improving it where necessary.

To fulfill these responsibilities, persons responsible for safety and health need to maintain direct contact with line and staff supervision.

The first-line supervisor is a key person in the accident prevention program. To the worker, the supervisor is management, because this is the management level that is closest to the people. The supervisor is also the prime communicator with employees on safety and health matters.

With guidance and help from the safety and health director, first-line supervisors should: Establish safe work practices and conditions.
- Enforce safety rules.
- Teach employees how to recognize hazards.
- Report all injuries and assure prompt treatment.
- Investigate the causes of all accidents and see that action to prevent recurrence is completed.

Accident prevention efforts and results should be included in the supervisor's performance evaluations.

Safety and Health Committees

Safety and health committees are found in almost every successful organization.
In union plants, joint labor-management safety and health committees are established in accordance with the labor agreement.

An efficient, smoothly operating committee is one in which management and employees are in agreement as to the limit of their duties and responsibilities. Also, labor and management <u>must</u> make every effort to carry out their obligations.

The committee organization will vary from plant to plant and from time to time. Much depends upon the size of the organization, the type of problems, and the smoothness and character of employee relations.

The basic function of a safety and health committee is to create and maintain an active interest in safety and health and to reduce accidents and occupational illnesses. The following duties are examples of those sometimes assigned to safety and health committees:

- Make a systematic inspection to discover and report potential health and safety hazards.
- Observe safety practices and procedures of the workforce.
- Review accident reports and corrective measures. Attempt to contribute a positive attitude toward safety and health.
- Listen to employees' concerns about safety and health matters.

Small Plant Organization

Active management and control of the small plant safety and health program may be vested in the chief executive, general manager, or in an experienced and qualified supervisor who has both authority and status.
There are several advantages inherent in small-scale operations, such as closer contact with the working force, more general acquaintance with the problems of the whole plant, and, frequently, less labor turnover. However, it may be difficult to justify a full-time safety and health professional, physician, nurse, or other medical services.

Scattered Operations

Organizations with operations in scattered locations and that require relatively few employees, such as some construction projects, face special inherent problems of organization. Their operations may be seasonal or intermittent and there may not be a sufficiently stable working force to operate committees effectively. The local manager, therefore, needs to adapt the safety and health program to the local conditions.

III. MAINTENANCE OF SAFE AND HEALTHY WORKING CONDITIONS

Inspection of Work Areas

Inspection of work areas can locate hazards and potential hazards which can adversely affect safety and health. Safety inspections are one of the principal means of locating accident sources. Removal of these hazards can lead to substantially improved accident prevention. In promptly correcting work conditions, management demonstrates to the employees its interest and sincerity in accident prevention.

Safety inspections should not be conducted primarily to find how many things are wrong, but rather to determine if everything is satisfactory. The whole purpose should be one of helpfulness in discovering conditions which, if corrected, will bring the plant up to accepted and approved standards.

It is advisable to schedule periodic inspections for the entire facility. Equipment or operations which present the greatest hazards should be inspected more frequently. Such inspections may be made monthly, semi-annually, annually, or at other suitable Intervals. Some types of equipment, such as elevators, boilers, unfired pressure vessels, and fire extinguishing equipment, are required by law to be inspected at specific, regular intervals. Chains, cables, ropes, and other equipment subject to severe strain in handling heavy materials should be inspected at specified intervals. A careful record should be kept of each inspection.

Fire Inspections

One of the hazards having the greatest effect on an industrial plant is fire. Consequently, a system should be set up for periodic inspections of all types of fire protective equipment. Such inspections should include water tanks, sprinkler systems, standpipes, hose, fire plugs, extinguishers, and all other equipment used for fire protection. The schedule of inspections should be closely followed and an accurate record maintained.

Health Surveys

Whenever there is a suspected health hazard, a special inspection should be made to determine the extent of the hazard and the precautions or mechanical safeguarding needed to provide and maintain safe conditions. The services of an industrial hygienist may be needed. Physical examinations should be made of employees exposed to occupational health hazards.

Job Safety Analysis

Job safety analysis (JSA) is a procedure used to review job practices and uncover hazards that may be present. It is one of the first steps in hazard and accident analysis and in safety training. Supervisors and employees in completing the JSA learn more about the job. Study of the JSA will suggest ways for improvement of the job methods, resulting in better work procedures and fewer accidents. A JSA is often kept near a machine so that an operator can review it at any time, especially when starting a new job.

IV. ESTABLISHMENT OF SAFETY AND HEALTH TRAINING

Employees

Effective safety training for all employees is an essential part of any successful accident and illness prevention program. New employees need primary safety orientation to provide a base for future attitude development. They should be taught the specific work practices necessary for their jobs. Job hazards should be identified and proper controls and procedures explained. In beginning on a job, the new employee must be given adequate supervision to assure that the new employee gets started safely.

Supervisors

Supervisors must be trained in all areas of their safety responsibilities, such as hazard identification, job safety analysis, job instruction training, accident investigations, and human relations. Training must be updated whenever processes or operations change. Subjects for training should be related to accident experience.

Job Instruction Training

Job instruction training (JIT), the procedure for teaching a person how to perform a particular job, is accepted as one of the teaching tools in a quality instruction program.

V. ACCIDENT RECORD/DATA COLLECTION SYSTEM

Good recordkeeping is the foundation of a sound approach to occupational safety and health.

Records

Records of accidents and injuries are essential to efficient and successful safety programs, just as records of production, costs, sales, and profits and losses, are essential to the efficient and successful operation of a business. Records supply the information necessary to transform haphazard, costly, ineffective safety work into a planned safety program that controls both conditions and costs.

It is legally required that the company keep proper accident/illness records. It is highly desirable to establish a system for recording all accidents, not just those involving injuries. What may cause a property-damage-only accident today can be the cause of tomorrow's serious-injury accident.

To reduce the possibility of serious complications following a minor accident, there should be a system for reporting all injuries, no matter how trivial, so that prompt first aid treatment can be given and the accident investigated.

Accident Investigation

It is obvious that every accident that occurs should be thoroughly investigated as soon as possible to find its cause and to prevent a recurrence. In addition to accident prevention, other benefits include cost reduction (both the direct and the more sizable indirect costs), continuation of operations or activities without disruption, and the maintenance of good employee morale with its frequently realized higher productivity and fewer work problems.

The important record is the accident investigation report. Every accident should be thoroughly investigated by the immediate supervisor, or depending on its severity, by an accident fact-finding committee appointed by top management.

During the investigation, special inspection of the accident scene is essential. The accident investigation identifies what action should be taken and what improvements are needed to prevent similar accidents occurring in the future. It also documents the facts for use in instances of compensation and litigation.

Accident Analysis

Analyzing accident records will provide convenient and systematized warning. Causal data should be available from the accident report, including such items as the type of injury and body part; general cause, such as unsafe act and/or unsafe conditions; and specific causes such as caught in, contact with, fall from, overexertion, struck by, or struck against. This detail should be analyzed and preventive countermeasures developed.

In the United States, injury and occupational illness records are required by the Occupational Safety and Health Act and specific record requirements are published by OSHA. Injuries and illnesses are recorded separately, with three different categories, as follows:

1. FATALITIES, regardless of the time between the injury and death, or the length of the illness;

2. LOST WORKDAY CASES, other than fatalities, that result in lost workdays;

3. NONFATAL CASES WITHOUT LOST WORKDAYS, which result in transfer to another job or termination of employment, or require medical treatment, or involve loss of consciousness or restriction of work or motion. This category also includes any diagnosed occupational illnesses that are reported to the employer, but are not classified as fatalities or lost workday cases.

Rates

Incidence rates, which relate the total number of injuries and illnesses per category to total employee-hours worked, can be computed from the formula:

$$\frac{\text{Number of injuries and illnesses} \times 200{,}000}{\text{Total hours worked by all employees during the period}}$$

The resulting rate will be expressed as incidents per 100 employees. These rates can be computed by operation within the plant to determine those areas with the most injuries. They can be used to detect incident rate trends within the plant. They can be used to compare your plant with similar plants or your industry.

Countermeasures

Once the accident and/or health hazard(s) has been identified and evaluated, then corrective action must be taken. In general, and in order of effectiveness, the items below should be considered.
- Change the system or machines, method, process, etc., to eliminate the hazard.
- Control the hazard by enclosing, guarding, etc.
- Train employees to increase awareness and to follow
- safe job procedures. Prescribe approved personal protection equipment, etc.

VI. HEALTH, MEDICAL, AND FIRST AID SYSTEMS

Health Services

Occupational health services deal with both the person and the work environment. A comprehensive health program requires (a) concern with all aspects of the work environment that may harm an individual, and (b) a constructive approach to industrial production problems through medical supervision of the employee's health.
The program should be supervised by a physician interested in industrial employees and qualified in industrial medicine. To be effective, the program needs certain medical and first aid facilities, a necessary staff, and the full cooperation of management.

Medical

Preplacement examinations should be conducted to determine and record the physical condition of the prospective worker so that the employee can be assigned to a suitable job. The individual capabilities should meet or exceed the job requirements. Safety must be a factor in the "employee-job fit." Periodic examinations of all employees are sometimes necessary.

First Aid

First aid is an important part of a safety and health program. Immediate, temporary treatment by a qualified individual should be available in the case of accident or sudden illness before the services of a physician can be secured (if they are needed).

VII. ACCEPTANCE OF PERSONAL ACCOUNTABILITY BY EMPLOYEE

Employees make many contributions to the accident prevention programs through the safety suggestions they make and the safety activities in which they participate. But above all, each employee must be trained to work safely and to accept responsibility for his or her own safe work practices. A high degree of employee pride should be developed in the safety record along with the motivation to maintain and improve that record.
To be effective, a program for maintaining interest in safety and health must be based on employee needs. Such activities as contests, drawings, family affairs and award presentations, and the like, serve to reinforce and communicate the safety and health program to the employees. Safety and health programs are a continuing activity, not a one-shot project.

SUMMARY

Successful safety and health programs have distinguishing characteristics. These include:

1. Strong management commitment to safety and health that is shown by various actions reflecting management's support and involvement in activities.
2. Close contact and interaction between workers, supervisors and management enabling open communications on safety and health as well as other job-related matters.
3. Training practices emphasizing early indoctrination and follow-up instruction in job-safety procedures.
4. Evidence of added features of variations in conventional safety and health practices serving to enhance their effectiveness.

Made in the USA
Middletown, DE
24 February 2025